More Random Walks in Science

Also by Robert L Weber, published by
The Institute of Physics

A Random Walk in Science

Pioneers of Science

More Random Walks in Science

An anthology compiled by
Robert L Weber

The Institute of Physics

Bristol and London

Published by The Institute of Physics
Techno House, Redcliffe Way, Bristol BS1 6NX
and 47 Belgrave Square, London SW1X 8QX

British Library Cataloguing in Publication Data
Weber, Robert L.
   More random walks in science.
   1. Science—Anecdotes, facetiae, satire, etc.
   I. Title
   502'.07   Q167

ISBN 0–85498–040–7

First published 1982
Reprinted 1984

Text set in Baskerville, Bodoni and Times Roman
Typeset, printed and bound in Great Britain by Pitman Press Ltd,
Bath

To Marion, Rob, Karen, Meredith and Ruth

Contents

x

xi

xiii

Preface

This is an anthology of humour and anecdote related to the sciences. It is intended for casual reading. Its goal is not analysis but rather enjoyment. Since, as Leo Rosten so rightly said, 'humour is the affectionate communication of insight' this Random Walk may lead you to a warmer feeling toward individual scientists and respect for the ingenuity and cheerfulness of the scientific spirit. I hope so.

A Random Walk in Science published in 1973 won numerous friends, many of whom have made contributions or suggested sources of pieces for this new book. In it the humour of physics receives most attention but you will also find irreverent quips and anecdotes about academic life in research, physics, mathematics, astronomy, space science, cosmology, chemistry, life sciences and earth sciences. In fact the book is roughly divided into sections corresponding to these topics. Those who originated these pieces or compiled and edited this anthology welcome your enjoyment of the lighthearted aspect of science.

The Institute of Physics and I thank all those who submitted articles and gave permission to publish them in this book. Detailed acknowledgments are given at the back of the book and in the marginal notes to articles. I would also like to thank Neville Hankins, Helen Gowie and Isobel Falconer of The Institute of Physics for their help in editing and producing *More Random Walks in Science*.

ROBERT L WEBER

xv

Splitting the infinitive

CHARLES R LANGMUIR

Contributed by
Donald Cooksey.

One of the most closely guarded secrets of the era can now be told, how an anonymous group of grammarians, working in secrecy in a remote section of the country, have finally succeeded in splitting the infinitive.

The so-called Bronx Project got under way in 1943, with the installation of a huge infinitron specially constructed for the job by California Tech philologists. Though the exact details are still withheld for reasons of security, it is possible to describe the general process.

From a stockpile of fissionable gerunds, encased in leaden clichés to prevent radioactivity, a suitable subject is withdrawn and placed in the infinitron together with a small amount of syntax. All this material must be handled with great care as the slightest slip may lead to a painful solecism. Once inside the apparatus, the gerund is whirled about at a great speed, meanwhile being bombarded by small participles. A man with a gender counter stands always ready to warn the others if the alpha-betical rays are released in such high quantities as to render the scientists neuter.

The effect of the bombardment is to dissociate the whirling parts of speech from one another until at length an infinitive splits off from its gerund and is ejected from the machine. It is picked up gingerly with a pair of hanging clauses and plunged into a bath of pleonasm. When it cools, it is ready for use.

The question is often asked: 'Can other countries likewise split the infinitive?' I think we can safely answer 'No.' Though it is true that Russia, for one, is known to have large supplies of thesaurus hidden away behind the Plural Mountains, it is doubtful if the Russians possess the scientific technique. They have the infinitive but not the knowhow.

And that is something on which to congratulate our own brave pioneers in the field of grammatical research. Once it was thought that the infinitive could never be split—at least not without terrible repercussions. We have shown that it is quite possible, given the necessary skill and courage, to unquestionably and without the slightest shadow of a doubt accomplish this modern miracle.

Broken English

H B G CASIMIR

Condensed from *Journal of Jocular Physics* vol. III (7 October 1955). The journal was published at intervals by The Institute for Theoretical Physics, Copenhagen, to celebrate the birthdays of Niels Bohr.

There exists today a universal language that is spoken and understood almost everywhere: it is Broken English. I am not referring to Pidgin English, a highly formalized and restricted branch of BE, but to the much more general language that is used by waiters in Hawaii, prostitutes in Paris and ambassadors in Washington, by businessmen from Buenos Aires, by scientists at international meetings and by dirty-postcard peddlers in Greece, in short by honourable people like myself all over the world.

One way of regarding Broken English is to consider it as a more or less successful attempt to speak correct English, but that is a pedantic, school-masterish point of view that moreover threatens to stultify the speaker of BE and to deprive his language of much of its primeval vigour. It is about time that Broken English is regarded as a language in its own right. It is then found that BE is a language of inexhaustible resources; rich, flexible, and with an almost unlimited freedom. In the following I shall try to establish some of the fundamental principles of BE in the hope that others, more qualified than myself, will take up the subject and help to secure for it the prominent place in linguistics to which it is justly entitled.

PHONETICS

The immense richness of BE becomes at once evident if we try to represent its sounds. Two short lines of keywords (44 in all) at the bottom of a page in a 25 cent Merriam–Webster are a sufficient clue to the pronunciation of standard American, and the famous pronouncing dictionary of Jones has only 35 keywords. Compare these pedestrian figures with the wealth of sounds current in BE. The whole international phonetic alphabet is hardly sufficient to meet the case. Take one simple letter like *r*. It may sound like an Italian *r* beautifully rolling on the tip of the tongue, like a guttural Parisian *r* or like no *r* at all. In this last case the speaker usually suffers from the illusion that he speaks pure Oxford English. Similarly *th* may sound as a more or less aspirated *d* or *t* or as a simple *z* and sometimes (especially in the case of Greeks) almost like *th*. Then there are elements entirely foreign to English like the Swedish musical accent and the Danish glottal stop (some people pretend that the glottal stop is hard to pronounce but that is nonsense; it is very easy in itself and only gets difficult if you try to put it into a word).

But even more important is the principle of free choice. It is well known that the combination *ough* can represent at least five different sounds. The educated speaker of BE is well aware of this fact but whereas the speaker of Standard English can only use *one* pronunciation in *one* word, the speaker of BE is at complete liberty. Some speakers make their choice once for all: they decide that they are going to pronounce doughnut as duffnut and stick to it. Others may use their freedom in a more subtle way and say duffnut or dunut depending on the hour of the day or the weather. Still others create distinctions and say dunut when referring to pastry but downut (like in plow) when referring to a circular discharge tube used in modern physical apparatus.

Then there is the accent. In Standard English this is a queer business. During the development of the English language the accent had a tendency to move to the front of a word but it has not gone all the way and it has shown a curious inclination to linger on the more irrelevant and meaningless syllables of a word. Words like barometer and turbidity will illustrate the point. Whether this is one more example of the traditional British sympathy for the underdog I do not know, but the result is baffling and to the convinced speaker of BE the realization that he has nothing to do with these weird intricacies comes as a great relief. The *dogmatists* will use their freedom by putting the accent always on the first syllable whereas the *rationalists* will stress what seems to them the most important syllable. The *quixotics* try to imitate Standard English. This is obviously impossible but the result has sometimes a certain slightly pathetic charm.

GRAMMAR

Much of what is said about phonetics applies to grammar too. Again a great richness, again a principle of free choice. The gain in power of expression that can be derived from, for example, a judicious use of the article is impressive. If a man invites you to a party it may very well turn out to be a dull

3

show, but if he says, 'Today we will have party and shall drink the cocktail,' you can almost be certain that you are in for a lively time. Changing the sequence of words gives new flavour to old sayings—'this is the moment when the frog into the water jumps,' one of my teachers used to say at the critical spot in a mathematical proof.

VOCABULARY

Also here there is a great freedom. Of course complete Humptydumptyism is impossible, but BE is the closest approach compatible with a measure of understandability. (To explain the term Humptydumptyism: 'The question is,' said Alice, 'whether you can make words mean different things.' 'The question is,' said Humpty Dumpty, 'Which is to be master, that is all.') It is characteristic of the genius of Lewis Carroll that he, who was by birth and breeding excluded from obtaining a mastery of Broken English, came by sheer artistic intuition so close to one of its basic principles.

Notwithstanding the great liberty in the use of words there is one case where all speakers of BE seem to agree: in Broken English, Broken English is called just 'English'.

IDIOM

It is remarkable how old and trite sayings in any ordinary language may acquire new glamour when translated into BE. The only danger is that one may unintentionally come to use an existing English proverb. 'Who burns his buttocks must sit on the blisters' sounds all right to me, but heaven knows whether a similar saying does not exist in Standard English.

I am afraid that this very short survey will have to do for the present. But I have still two important remarks to make. Firstly, in view of the stupendous wealth of BE, it will at once be evident how completely ridiculous, ludicrous, preposterous, and ill-advised are the attempts to introduce for use by foreigners a so-called Basic English, a language not richer but even poorer than Standard English. Secondly, it is often stated that at the age of 16 or so one loses the faculty to learn English correctly. This again is entirely wrong: nothing of any importance is lost; what is gained is the faculty to create one's own brand of Broken English.

4

NPL haiku

From *NPL News* **204** 20(21 April 1967).

'The haiku form is simple—a verse of 17 syllables, divided into three lines of five, seven, and five syllables respectively. The Western ear should note that the metrical unit is the syllable (Japanese is a syllabic language) and not, as in Western prosody, the foot composed of two or more syllables. The form of 17 syllables is not chance; it derives from the traditional view of Japanese linguistic philosophy that 17 syllables is the optimum length of human speech to be delivered clearly and coherently in one breathing.'

In asking its readers to compose haikus for a competition, the [Manchester] *Guardian* quoted a modern exponent as saying 'Haiku presents a kind of knowledge different from the scientific, offering intuitive insight into the deepest levels of meaning.' Not quite believing that our scientists were so cut off from the deepest levels of meaning, we asked a few to try their hands at writing in haiku form on a subject to do with NPL.

Editor advised
Preserve anonymity
Or no more copy

With wise levity
It is best to meet problems
Of great gravity

Our deepest thoughts are
But scratches at some door that
Nature keeps ajar

Celestial masers
May possibly simulate
Peculiar quasars

Hope[1] springs eternal
To tap our feeble talents
Love'll find a way

[[1]Miss Hope Lovell, until her retirement in August 1970, was Editor of *NPL News*.]

Designed originally for preliminary studies in the field of four-dimensional stress analysis, this framework is now being used to form the basis of a new type of electronic package which will of course be able to accommodate both real and imaginary components (P L Kirby).

Federalese

G R HICKS

From *Word Study* (Springfield, MA: G & C Merriam) May 1956.

Channels—*The trail left by inter-office memos.*

Consultant (or Expert)—*Any ordinary guy more than 50 miles from home.*

To activate—*To make carbons and add more names to the memo.*

To implement a program—*To hire more people and expand the office.*

To clarify—*To fill in the background with so many details that the foreground goes underground.*

We are making a survey—*We need more time to think of an answer.*

Note and initial—*Let's spread the responsibility for this.*

Will advise you in due course—*If we figure it out, we'll let you know.*

Referred to a higher authority—*Pigeonholed in a more sumptuous office.*

Research work—*Hunting for the guy who moved the files.*

Further substantiating data is necessary—*We've lost your stuff. Send it in again.*

We're exploring the problem—*Don't get impatient. We'll think of something.*

Readability ratings

From *The New York Times* (11 August 1976).

Using the Rudolph Flesh readability scale—rating by length of sentences and words—[Herbert S Denenberg, former Pennsylvania Insurance Commissioner,] scored auto policies as 10 and lower, compared with 80.7 for Joe Gargiola's *Baseball is a Funny Game* and 17.72 for Einstein's *Meaning of Relativity*. 'It's easier to understand Einstein's theory of relativity than the average auto policy' he said.

From *Applied Optics* **8** 273 (1969).

God grant that no one else has done
The work I want to do,
Then give me the wit to write it up
In decent English, too.

A note on the game of refereeing

J M CHAMBERS† and AGNES M HERZBERG‡

From *Applied Statistics* **XVII** No 3 (1968).

†*Bell Telephone Laboratories, Murray Hill, New Jersey*
‡*Imperial College, London*

SUMMARY

The game of refereeing is described. Some tactics are outlined and examples are given. The present state of the game is assessed.

I. INTRODUCTION

This paper defines the game of refereeing. While not a new game by any means, it is now played more widely than ever before, as the volume of current journals demonstrates. Here we outline the game and suggest some of the more effective tactics. The examples are drawn from the field of statistics, but the reader may easily supply examples from many other areas.

The version of the game outlined here represents a reasonably high standard of rigour. While it would be foolish to assert that the game is not frequently played under less restrictive conditions, we are convinced that it can only lose in subtlety, intellectual interest and artistic scope by such relaxations.

The game is played between two teams, here called author and referee. The former consists of one or more co-operating players. We treat these as a single player. In the simple or univariate game there is a single referee who plays against the author. This is the case treated in this paper. Discussion of the p-referee case for $p \geqslant 2$ involves no basic alteration in the model or rules.

The definition of the objective for the players and the optimum being sought have been the subject of considerable discussion. It is agreed that the author's objective is to have his paper published, and that extra points accrue for the publication of a particularly worthless submission. Likewise the referee's minimal objective is to have the paper refused and extra credit is obtained if the paper was a major contribution to the field. Some consider that the referee may attain a higher optimum if, in addition to having the paper refused, he reduces the author to a 'nervous pulp' (a term introduced in connection with another game, Conference-manship). Still greater success can be claimed if the author gives up serious work altogether, say by joining an operations research firm. Similar higher-level goals might be devised for the author; for example, taking up all the time the referee would normally spend on his own research.

Play opens with the *submission* of the paper by the author. At this point the editor of the journal intervenes to select the opposing player(s). (We consider the editor as a neutral umpire, deferring the question of the play of editor versus author and/or referee to a subsequent paper now in progress by these authors.) The next move is by the referee. Without loss of generality we call this move the *refusal*. This may be followed by a further submission, a further refusal and so on, until one or other player concedes defeat.

These are the basic rules of play. In Section 2 we give some of the more useful tactics which may be employed by the players.

2. TACTICS FOR THE GAME

Note: The tactics given below are not intended to be exhaustive. They are given as examples to help the novice player. Obviously the tactics may be used even more effectively in combination.

2.1 Tactics for the Author

A1. Obscure-reference tactic. Here the author refers to a paper in an obscure journal; for example, *Journal of the Indian Statistical Association, Scientia Sinica*, or *Applied Statistics*. The paper has, preferably, a rather general title and, therefore, might possibly include the point which the author is trying to make, for instance, 'A result on limit theorems', 'A note on the analysis of variance', 'On statistical estimation', or more simply, 'An aspect of statistical theory'. It should be noted that, whether or not the paper is by the person who refers to it, a previous game of refereeing has taken place to get it published. Another way to employ this tactic is for the author to refer to a private communication from possibly a well-known statistician or to

9

his own or others' unpublished work. Unfortunately, this version of the tactic is not allowed by some journals.

A2. Wrong-reference tactic. The author refers to a paper which is not given or is incorrectly given in the list of references. This will infuriate a conscientious referee who insists on checking all references in an effort to show that the author has not read the literature properly. (The author must be careful in employing this tactic for some referees think that even a misprint in the references is reason enough for rejecting an otherwise worthy paper.)

A3. Prestige tactic. The author uses at least one reference to a well-known person to show that he is working in a 'good' field. It is usually possible to find such papers with general titles to cover a large variety of sins.

A4. Barrage tactic. The author sends in such a large number of papers to the same journal that the editor cannot cope with them and will, therefore, have to let some be published without proper refereeing. This prevents the referee from entering the game at all and thus the author is essentially playing a game of solitaire. The author could also send the same paper to various journals with a slight change in title, thereby playing more than one opponent at the same time. There is, of course, the possibility that the same referee may be chosen by more than one journal.

A5. Flattery-may-get-you-somewhere tactic. In the revision of the paper the author thanks the referee for his 'helpful comments' etc. This is very often employed against tactic R_5 by saying something to the effect that he (the author) 'agrees that he was not clear in the earlier version of the paper'.

A6. Anticipation tactic. Here the author attempts to disarm criticism either:

(a) by inserting flattering references to the work of all the more likely potential referees; or

(b) by writing papers jointly with all the experts in the field, thus making it impossible to find a referee.

(In (a) the author of a bibliography is at an advantage. In (b) the game becomes a game of solitaire.)

A7. Precedent tactic. Reference is made to a paper which although of very low quality was recently published in the same journal. The author implies that his work cannot be of lower quality than the previous paper. The danger, however, is that the editor may be only too aware that he should have rejected that paper and will act accordingly.

A8. Deliberate-mistake tactic. A deliberate, obvious and unimportant mistake can be inserted near the beginning of the paper. An inexperienced referee will use it to suggest rejecting the paper and then will be overruled by the editor. The author must be careful in employing this tactic since an experienced referee will use the mistake merely to suggest (without actually saying) that the whole paper is full of such mistakes.

2.2 Tactics for the Referee

R1. Obscure-reference tactic. The tactic *A1* may also be employed by the referee: for example, by suggesting with a reference to an obscure paper that the author's work is not original.

R2. Wrong-level tactic. No matter what degree of rigour the author uses, the referee replies by saying that it is not the correct one. For example, 'The author has stressed rigour to the detriment of clarity', 'The author's colloquial style is insufficiently rigorous', 'The author unfortunately tries to combine rigour with a colloquial style to the detriment of both'.

R3. Unsuitable-for-publication-in-this-journal tactic. This tactic is also known as the 'shirking-of-duty tactic'. As a last resort the referee says that the paper is unsuitable for publication in the journal in question, and makes a suggestion that it be submitted to another journal which is suitably insulting to the author. This then ends the game between these two particular opponents. The referee then hopes that the suitably insulting journal does not ask him to referee the paper.

R4. Shorten-paper tactic. In spite of the fact that more and more journals are publishing more and more times a year and that each issue must be of a respectable size, editors seem to prefer short papers. Therefore, the referee can always request that the paper be shortened. This usually gives the author a difficult task and will tend to prolong the game.

As a counter to tactic *A6*(a), the referee may suggest publication of only the sections containing flattering references to his own work.

R5. Deliberate-misunderstanding tactic. The referee deliberately questions something in the paper which he knows to be correct. This is a delaying tactic.

R6. Personal-knowledge tactic. The referee, knowing who the author is, questions points in the paper on which he knows that the author knows nothing. This makes the author nervous about what he has written. This tactic cannot be used in those journals (for example, *Psychometrika*) where both players remain unknown to each other.

R7. Standard-vs.-unstandardized-notation tactic. Whatever notation the author uses the referee replies that this should be changed to the standard notation; for example, Author: Let x and y represent the variables. Referee: Change x and y to a and b (or vice-versa).

R8. Scare tactic. In commenting on the author's paper the referee refers to a paper of his own 'in press' (the paper may or may not have been started). The title of the paper suggests it may include the author's work. With a nervous or inexperienced opponent this may terminate the game.

R9. Frustration tactic. Perhaps the most important of the referee's tactics is to do nothing and to ignore all correspondence about the paper; this is of course particularly

12

effective with handwritten manuscripts. The experienced referee will not trust his colleagues and will deposit the manuscript at the bank.

3. CONCLUSION

We have described the game as currently played. There are, however, many possible improvements. The range of strategies would be greatly extended if collusion among referees were allowed. Careful use of mutually contradictory requirements by different referees will help greatly to demoralize the author. Authors, on the other hand, would have a number of interesting new tactics if the name of the referee were given to them.

It must be acknowledged that the entire practice of refereemanship has declined in recent years. With the publication of more and more journals, and the issuing of present journals more frequently, the pressure for papers to fill them restricts the referee from rejecting as many acceptable papers as hitherto. Further, there are now so many papers which deserve rejection on their own merit that true skill in refereemanship is no longer in demand. Improvements in communication and in the work of librarians have made some of the tactics (e.g. $A1$ and $R1$) more difficult to apply.

However, the most insidious cause of this decline is the loss of the true savage refereeing spirit among the modern generation of players. We fear that too many participants have taken to heart the old adage, 'Referee as you would have others referee when you are writing.'

From *Applied Optics* **7** 1915 (1968). Writers in Britain today are not so much bothered by censors—editors are our bugbear. But, in commenting on them, few would choose the word *glutton*. Why should the poor devils not be gluttonous? During the post-prandial snooze some of our stuff might slip through unmangled.

13

Preparing scientific papers

N S HAILE

From *Nature* **268** 100 (1977).

A great deal of time is wasted because young scientists submit papers for publication in an unacceptable form. There are many good books on the market on the preparation of scientific papers, but few specific examples as to how an unacceptable manuscript can be transformed into one acceptable by leading scientific journals. By a lucky chance (and completely legally) I was able to obtain a copy of such a manuscript with the referees' comments and a model rewritten version, from the editorial files of a leading geological journal. I pass this on in the hope that it will be of value to authors in preparing papers for publication.

COLUMNAR ROCK STRUCTURES FROM AN ANTIQUE LAND
Referees' report: manuscript 19705B/76: P B Shelley

Manuscript as submitted:
Ozymandias[1] by P B Shelley[2]

I met a traveller[2] from an antique land[3]
Who said: Two[4] vast[5] and trunkless legs[6] of stone[7]
Stand in the desert.[8] Near them,[9] on the sand,[10]
Half sunk, a shattered visage lies, whose frown,
And wrinkled lip, and sneer of cold command,
Tell that its sculptor well those passions read
Which yet survive (stamped on these lifeless things),
The hand that mocked them and the heart that fed;[11]
And on the pedestal[12] these words appear:
'My name is Ozymandias, king of kings;
Look on my works, ye Mighty, and despair!'[13]
Nothing beside remains.[14] Round the decay
Of that colossal wreck, boundless[15] and bare,
The lone and level sands[16] stretch far away.

Referees' comments
[1]This title is quite inadequate. Includes no keywords. See suggestion below.
[2]Since this paper appears to be based on field observations by another geologist, we suggest joint authorship would be appropriate.

³Specify.

⁴This is the only quantitative statement!

⁵Not specific enough. The authors should give dimensions in SI units. (Unless 'vast' is a class in some sort of grade-scale, in which case the reference to this scale should be given.)

⁶Have alternative hypotheses been considered? Earth pillars? Basalt columns? Ant hills?

⁷Surely identification of rock type with appropriate analyses could be provided here?

⁸Give co-ordinates.

⁹Specify distance. A photograph (giving scale) would help here.

¹⁰Give granulometric analysis, and preferably some scanning electron microscope photographs of grain-surface textures. These don't actually prove anything, but are decorative and keep SEM operatives in employment.

¹¹This fanciful and speculative section could well be omitted.

¹²This is the first we have heard of a pedestal!

¹³While it may be worth-while to record the defacement of an interesting exposure, it is not necessary to quote the words. (Since they are in English, they are obviously of no archaeological interest. Presumably graffiti sprayed on by a tourist.)

¹⁴Rather dogmatic. Better: 'No other rock exposures *were observed*'.

¹⁵Inappropriate hyperbole. The approximate extent of the desert should be stated, if relevant.

¹⁶Unless this is a windless desert, surely a sandy desert should show dune formation? If actually level, perhaps in fact it is a stony desert?

General remarks Although some interesting observations are recorded, we cannot recommend publication in the present form. For one thing, the paper is far too short. For another, the authors have inexplicably left out any mention of plate tectonics! For the guidance of the authors, we give below a summary of the kind of re-written and expanded paper which might be acceptable. We have had to supply necessary missing data arbitrarily; it is, of course, up to the authors to substitute the correct data.

Suggested re-written manuscript (summary):

Twin limb-like basalt columns ('trunkless legs') near Wadi Al-Fazar, and their relationship to plate tectonics.
Ibn Batuta[1] and P B Shelley[2]

In a recent field trip to north Hadhramaut, the first author observed two stone leg-like columns 14.7 m high by 1.8 m in diameter (medium vast, ASTM grade-scale for trunkless legs) rising from sandy desert 12.5 km southwest of Wadi Al-Fazar (Grid 474 753). The rock is a tholeiitic basalt (table 1); 45 analyses by neutron activation technique show that it is much the same as any other tholeiitic basalt (table 2). A large boulder 6 m southeast of the columns has been identified as of the 'shattered visage' type according to the classification of Pettijohn (1948, page 72). Granulometric analysis of the surrounding sand shows it to be a multimodal leptokurtic slightly positively skewed fine sand with a slight but persistent smell of camel dung. Four hundred and seventy two scanning electron photomicrographs were taken of sand grains and forty are reproduced here; it is obvious from a glance that the grains have been derived from pre-cambrian anorthosite and have undergone four major glaciations, two subductions, and a prolonged dry spell. One grain shows unique lozenge-shaped impact pits and heart-like etching patterns which prove that it spent some time in upstate New York.

There is no particular reason to suppose that the columns do not mark the site of a former hotspot, mantle plume, triple junction, transform fault, or abduction zone (or perhaps all of these).

Keywords plate tectonics, subduction, obduction, hotspots, mantle plume, triple junction, transform fault, trunkless leg, shattered visage.

[1]School of Earth & Planetary Sciences, University of the Fertile Crescent.
[2]Formerly of University College, Oxford.

Even if I could be Shakespeare, I think I should still choose to be Faraday.

ALDOUS HUXLEY (1925)

Commandments for the writer of a dissertation

Adapted from
Physicists continue to laugh (Moscow: MIR Publishers)
1968.

[*Compiled by bored members of the Science Council during the defence of a dissertation; reproduced by grateful dissertation writers.*]

PREPARATION OF THE DISSERTATION

(1) Do not write at length. A dissertation is not *War and Peace* and you are not Leo Tolstoy. A bulky dissertation is to your opponents like a red rag to a bull.

(2) Do not write compactly. This attests either to great talent or to barrenness of the mind. Your opponents will not forgive you either one or the other.

(3) The title for the dissertation is as functional as a hat for a woman in summer.

(4) Observe the amount of 'for' and 'against' in selecting literature. When there is a lot of material 'against' in a dissertation, doubt of the integrity of your opinions sets in. If only data 'for' are introduced, your merits will be obscured.

(5) Do not tap classics of natural science on the shoulder.

(6) Do not put on airs. Do not think that all who surround you are fools and that you alone are intelligent. Avoid personal pronouns: replace 'I consider' with the modest 'evidently one can consider'.

(7) Check the quality of the dissertation on family members and colleagues. A normal dissertation must evoke involuntary yawning and subsequent sleep. Sections which provoke spasms of mirth or a feeling of oppressive unrest must be rewritten. Do not be glad if an unsophisticated listener says that everything is intelligible to him: this is a true indication that you will not be intelligible to a scholarly listener.

SELECTION OF OPPONENTS

(8) The opponent is the central figure in the defence.

(9) The ideal opponent must have a general notion about the subject of the dissertation but must not be a specialist on the given question. An opponent completely unacquainted with the question can render service as a devil's advocate praising precisely that which should be mildly criticized, whereas a specialist delves into details undesirable for public discussion.

(10) Avoid inviting as opponents young degree candidates and those who have just recently received their doctorates.

They have only just won their 'place in the sun' and are always glad to profit from a chance to show themselves off and discredit others. It is far more satisfactory to invite venerable, honoured practitioners of science because in old age we are all made if not better at any rate lazier.

(11) Try to make probable unofficial opponents into participants in the defence. For this turn to them for advice and thank them for valuable aid. Thereby you demonstrate your insignificance and their superiority. Thus you turn a foe into a person interested in the outcome of the defence; after all, who wants to come out against his own recommendations?

THE DEFENCE OF THE DISSERTATION

(12) There is no greater foe of a dissertation than the writer himself. The writer of a thesis is liable to have a distorted view of it. The natural law of this phenomenon, confirmed in nearly 100% of the cases, compels one to reckon with it. Taking this into consideration, repeat your speech (performance) over and over again at home.

(13) On the rostrum conduct yourself with decorum. Do not pick at your ears, do not twirl your pointer over the heads of the members of the praesidium, do not drink more than one glass of water, do not weep, do not blow your nose.

(14) If the report is written, do not make a speech [of it] but read it. On the other hand the mumbling of the dissertation writer causes resentment in the listeners. Try to speak in a monotone. The more that members of the Science Council sleep or daydream about personal affairs the quicker and more successfully will the defence progress.

(15) Illustrative material is very important. One can show off a quantity of factual material by using a projector. For this give orders to the operator: 'Show Curve No. 25.' 'Tables No. 8 to No. 24.' Of course, it is not obligatory to select the relevant material. The operator doesn't care what is shown and the very fact of the abundance of material captivates the listener. If you use tables, hang up a few more of them. It stands to reason that you will have to dwell on only a few. The remaining ones indicate a background of extensive experimental material.

18

(16) In your concluding words thank and bow, bow and thank. Strictly observe the necessary order of rank. Thank those absent less, thank those present more.

(17) After a successful defence arrange for a banquet.

When Wittgenstein submitted his *Tractatus* as his doctoral dissertation, G E Moore is said to have reported as an examiner: 'Mr Wittgenstein's thesis is a work of genius; but, be that as it may, it is certainly well up to the standard required for the Cambridge degree of Doctor of Philosophy.'

My German engineer, I think, is a fool. He thinks nothing empirical is knowable. I asked him to admit that there was not a rhinoceros in the room, but he wouldn't. I looked under all the desks without finding one but Wittgenstein remained unconvinced.

BERTRAND RUSSELL

From *A Mathematician's Miscellany* by John E Littlewood (London: Methuen) 1960.

SCHOOLMASTER: 'Suppose x is the number of sheep in the problem.'

PUPIL: 'But, Sir, suppose x is not the number of sheep.'

[*I asked Professor Wittgenstein was this not a profound philosophical joke, and he said it was.*]

Science is always wrong. It never solves a problem without creating ten more.

GEORGE BERNARD SHAW

Royalties: the Quatorze connection

MARQUE BAGSHAW

From *Bulletin* of the Graduate School, Pennsylvania State University (2 May 1977).
University Microfilms, the company that microfilms and sells to scholars copies of doctoral theses, recently announced that, beginning last year, they began paying royalties to authors of dissertations. If sales of a dissertation amount to $100 or more in a calendar year, the company says it will pay a royalty of 10% to the author, based on total sales for the year. Most dissertations sell at best only a few copies each, but with this incentive to turn out a marketable commodity, we soon may be seeing such come-on thesis titles as: 'Everything You Wanted to Know About *Lactobacillus Acidophilus* but Were Afraid to Ask'; 'Bill & Dottie & Sam & Sadie: Wordsworth and His Circle'; 'The Real Watergate: Designs for a High-pressure Check Valve'; 'Leaping Louis; or Capet's Last Caper (Eh, Maintenon?)'; etc.

Perish, but publish!

WILLIAM J MCGLOTHLIN

[*On the occasion of publishing an article after retirement.*]

From *AAUP Bulletin* (March 1978) p 25.
Old scholars retire, but never say die,
For print they eternally cherish.
They'll comb the Great Lib'ry up in the Sky,
And publish long after they perish.

Clarification

From *Applied Optics* **7** 19 (1968).
It is necessary for technical reasons that these warheads be stored upside down; that is, with the top at the bottom and the bottom at the top. In order that there may be no doubt as to which is the bottom and which is the top, it will be seen to that the bottom of each warhead immediately be labelled with the word TOP.

BRITISH ADMIRALTY

How to win arguments

V D LANDON

From *Proceedings of the IRE* **41** (9) 1188 (1953).

The first requirement in winning an argument is to start an argument. Fortunately this is one of the easiest parts of the process. For the benefit of the few who have trouble getting started, the following standard methods are listed:

(1) Make a loud aggressive statement belittling your friend's favourite hobby.

(2) Listen to your friend until he makes a positive statement of some sort. Contradict it flatly.

(3) Make one or more bold positive statements of doubtful validity. The subject doesn't matter, but politics, the weather, hobbies and girls are favourites. Or a technical subject will do, provided neither you nor your friend are well versed in the speciality discussed.

THE TECHNIQUES OF ARGUMENTATION

The circular path
One of the favourite artifices of expert arguliers is constant repetition. Perfect a pat argument of about a hundred words and repeat it time after time with as little variation as possible. This will wear your opponent down until he gives up in disgust.

The break-in or yeabut
Constant interruption is another favourite strategem. To be most effective, listen intently until it appears that your opponent is about to make an effective point, then interrupt in a loud voice and go into your own routine. Break-ins should usually be preceded by 'yeabut' whether you agree with any part of your opponent's argument or not.

The crescendo
This is a contrivance which is very effective in the hands of a master craftsman. In executing this manoeuvre, proceed with your standard routine until your opponent sees fit to interrupt. When this occurs pay absolutely no attention to what he says other than gradually to increase the volume of your own voice. This subtle ruse is usually met by a similar crescendo on the part of the opponent. It is a sure sign of the expert to be

able to shout back at your opponent so loudly and continuously as not to be able to hear a word he is saying.

The stutter

A disadvantage of the crescendo is that while it obliterates your opponent's argument it does not succeed in getting your own argument heard. For this reason the stutter is preferred by many experts. The stutter consists of a loud repetition of the first syllable of your proposed commentary with or without the crescendo. As soon as your opponent realizes he is being drowned out he will usually subside and you can proceed. It is an awesome spectacle the way a truly great master of the art can alternate the stutter and the crescendo to beat down all opposition. The stutter is a particularly effective chisel enabling you to break-in on your opponent's routine.

The use of invective

A further help is a large vocabulary of sacrilegious words and phrases skilfully adapted to describe your opponent's intelligence and ancestors. While this expedient is very effective, its advocates find it desirable to have some acquaintance with the art of pugilism.

The syllogism

To fall back on cold logic is a sign of weakness. It should never be indulged in unless you want to lose your reputation as a prize pain-in-the-neck.

The acceptance of a scientific idea

From *Journal of Genetics* **58** 464 (1963).

Professor J B S Haldane has described the normal process of acceptance of a scientific idea, in four stages:

(i)   this is worthless nonsense;
(ii)   this is an interesting, but perverse, point of view;
(iii)   this is true, but quite unimportant;
(iv)   I always said so.

On the reading of scientific papers
Audience enemies numbers I to VI

EUGENE F DUBOIS

Condensed from
Science **95** 273–4
(1942).

(I) The Mumbler, who drops his voice to emphasize important points or else talks to the lantern screen instead of to the audience.

(II) The Slide Crowder, who packs his slides with typewritten data and shows too many slides.

(III) The Time Ignorer, who talks beyond the limit specified in the programme or justified by common courtesy.

(IV) The Sloppy Arranger, who jumbles his material.

(V) The Lean Producer, who has poor material.

(VI) The Grasping Discussor, who when he gets talking stays talking.

Quoted by
R B Lindsay and
H Margenau in
*Foundations of
Physics* (Chichester:
John Wiley).

Mr Spencer in the course of his remarks regretted that so many members of the Section were in the habit of employing the word Force in a sense too limited and definite to be of any use in a complete theory of evolution. He had himself always been careful to preserve that largeness of meaning which was too often lost sight of in elementary works. This was best done by using the word sometimes in one sense and sometimes in another and in this way he trusted that he had made the word occupy a sufficiently large field of thought.

[*James Clerk Maxwell's humorous commentary on Herbert Spencer's appearance before the British Association in Belfast in 1874.*]

The most prominent requisite to a lecturer, though perhaps not really the most important, is a good delivery; for though to all true philosophers science and nature will have charms innumerable in every dress, yet I am sorry to say that the generality of mankind cannot accompany us one short hour unless the path is strewed with flowers.

MICHAEL FARADAY *Advice to Lecturers*

23

The use of small dogs in physics teaching

A W S TARRANT

From *Physics Bulletin* (December 1973) p 731.

Dogs are rarely met within university lecture rooms, and whenever they do appear there, their presence is generally not very welcome and is of short duration. My small four-legged lecture assistant, 'Rushton', provides an exception to this rule—he is portrayed in these pages. Rushton's origins are not quite clear, but there is no doubt that he was originally modelled on one of my children's toy dogs. His first appearance at a certain well known university in the southern part of England was confined to some rather pithy comments on various aspects of university life (figure 1). But he first really appeared before the public when I pressed him into service as my lecture assistant.

1

UNDERGRADUATE TEACHING

I should perhaps say that the lectures were a little bit unusual. At the University of Surrey we have developed common combined courses during the first two terms of some undergraduate courses, and one of these was the combined course for students of all the biological sciences (biochemistry, human biology, microbiology and nutrition). Now these students, both during their course and in later life, have to make quite extensive use of sophisticated instruments: spectrophotometers, oscilloscopes, electrocardiographs, pH meters and so on. Unfortunately, their knowledge of physics is not too good because it often is not treated terribly seriously by sixth formers heading for the life sciences, and there are a few students in these courses who have done no physics at all.

It was therefore decided to include in the course a set of lectures on 'measurement science' to enable these students to have a better idea of what instruments can and cannot do.

Unofficially the course was subtitled 'How not to make a fool of yourself with sophisticated instruments.' I had the pleasure of devising the syllabus and doing all the lecturing, and one day it occurred to me that Rushton might help.

2

He first appeared as he is in figure 2. Like many lecture assistants, he likes doing a bit of his own research, and here he is investigating an ant (or is it a spider?). Now this is research at its lowest level, without any instruments at all. All he can find out about his ant is that it smells of formic acid, and if you put your paw on it, it wriggles out through the fur. Rushton thereupon sought the help of instruments, and in figure 3 you see him setting up a microscope and camera to do a photo-micrograph. Unfortunately though, the ant has got fed up with the heat coming up through the condenser and has pushed off, so Rushton will discover when he develops his picture that he has got a completely blank plate. In other words, the instrument has not told him an untrue story—the stage was in fact blank—but it was not the story Rushton was expecting. This illustrates a most important point: when you use a complicated instrument, are you measuring what you think you are measuring?

3

Rushton also made himself useful by recalling simple experiments which the students had done or seen at some time in the distant past. It is a lot easier to have a picture of Rushton doing the experiment on a lantern slide than actually to do it in minimal time on the lecture bench. He made his mark, however, in helping to explain more complicated instruments, and particularly, closed-loop control systems.

Closed-loop control systems are of particular importance in this field, not only because sophisticated instruments make very extensive use of them, but because many chemical and biological systems (including *Homo sapiens*) also involve them. To the average physicist or mathematician, it is child's play to formulate their operation (as long as they remain linear) but to this audience the use of a second-order linear differential equation would not only be unfamiliar; it would be a waste of time.

In my opinion the basic idea of closed-loop control is best described by referring to actual instruments. The ones that I use are the high-speed recording potentiometer and the voltage stabilizer. We asked Rushton to set up a voltage stabilizer, given a DC supply main, and he produced the set-up in figure 4.

4

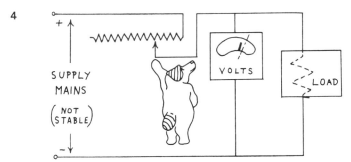

SUPPLY MAINS

$\left(\begin{smallmatrix} \text{NOT} \\ \text{STABLE} \end{smallmatrix}\right)$

VOLTS

LOAD

Most electrical loads are resistive, so he drew the load as though it was a kind of resistance. Current is fed from the DC main through the rheostat to the load and the voltage is indicated on the meter. Rushton has drawn a chalk mark at the desired voltage on the meter and he just watches the meter; if the load voltage falls he pushes the rheostat slider to the right a bit, thereby increasing the current and thus the voltage, to restore the status quo, and if the voltage goes too high he pushes it to the left a bit. (Well may you laugh, but in my first employment in a famous laboratory, that was exactly how we controlled the voltage on our standard lamps.)

Rushton's stabilizer has all the essential features of a closed-loop system. The voltmeter provides the measuring element, the chalk mark provides the reference element. Rushton himself is the amplifying element and the rheostat provides the controlling element. Rushton's job in this set-up is entirely automatic and could, of course, be done more reliably by a lot of electronics, as indeed it is in the typical output-controlled stabilizer. But by this simple lash-up Rushton is able to convey the basic idea of closed-loop control in a readily understandable way.

Rushton's stabilizer can demonstrate other features of closed-loop control. Consider the effect of a sudden large increase in the mains voltage (figure 5). The voltage change has been too fast for him so that the load is suffering a severe overvoltage; the load is just bursting into flames. Rushton is hurling over his rheostat, only to find that he has overdone it (figure 6). To restore the desired voltage as soon as possible he

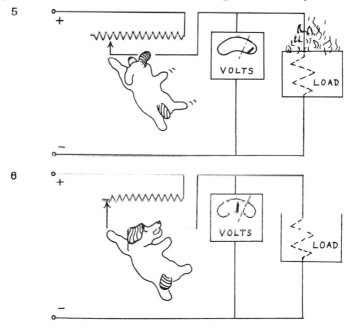

7

8

9

10

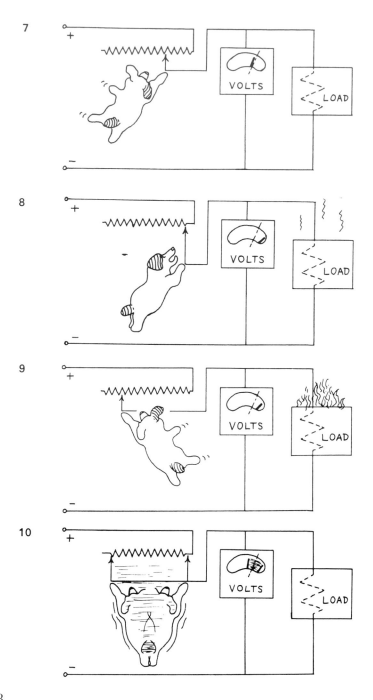

thereupon hurls it back again (figure 7) to discover that he has once again overvolted the load (figure 8), so he hurls it over again (figure 9). But this is where we came in; the situation is the same as it was in figure 5. Consequently, Rushton will go round and round this cycle faster and faster until he reaches the situation of figure 10, with both he and the voltmeter in continuous oscillation. The important thing is that Rushton is correcting *too hard, too late*. In fact, any closed-loop system where the correction is too hard and too late will go into continuous oscillation if there is not enough damping (voltage stabilizers included).

Rushton is well able to make this point in non-mathematical terms, and in my lectures he is followed up by a demonstration of a recording potentiometer behaving in the same way. Of course, when Rushton helps me in this way it is necessary to be *very* careful in recognizing his limitations. The use of a little fantasy certainly does help students for whom this is ancillary material; one biology student told me that 'she could understand it as long as Rushton was in the picture.' But I do not think that I had better ask him to explain entropy—he nearly got elected to a senior committee at his university the other day, and I had quite enough trouble talking my way out of that one.

Teaching

PICASSO

From *Life with Picasso* by Françoise Gelot and Carlton Lake (New York: McGraw-Hill) 1965, p 66.

So how do you go about teaching them something new? By mixing what they know with what they don't know. Then, when they see vaguely in their fog something they recognize, they think, 'Ah, I know that.' And it's just one more step to, 'Ah, I know the whole thing.' And their mind thrusts forward into the unknown and they begin to recognize what they didn't know before and they increase their powers of understanding.

Teacher evaluation

John Gauss in *Phi Delta Kappan* **43** (4) (January 1962). Reprinted in *The Saturday Review* (21 July 1962) and *The Physics Teacher* (April 1966).

TEACHER: Socrates

A. PERSONAL QUALIFICATIONS

	Rating (high to low)					Comments
	1	2	3	4	5	
Personal appearance	☐	☐	☐	☐	x	*Dresses in an old sheet draped about his body*
Self-confidence	☐	☐	☐	☐	x	*Not sure of himself—always asking questions*
Use of English	☐	☐	☐	x	☐	*Speaks with a heavy Greek accent*
Adaptability	☐	☐	☐	☐	x	*Prone to suicide by poison when under duress*

B. CLASS MANAGEMENT

	1	2	3	4	5	Comments
Organization	☐	☐	☐	☐	x	*Does not keep a seating chart*
Room appearance	☐	☐	☐	x	☐	*Does not have eye-catching bulletin boards*
Utilization of supplies	x	☐	☐	☐	☐	*Does not use supplies*

C. TEACHER–PUPIL RELATIONSHIPS

	1	2	3	4	5	Comments
Tact and consideration	☐	☐	☐	☐	x	*Places students in embarrassing situation by asking questions*
Attitude of class	☐	x	☐	☐	☐	*Class is friendly*

D. TECHNIQUES OF TEACHING

	1	2	3	4	5	Comments
Daily preparation	☐	☐	☐	☐	x	*Does not keep daily lesson plans*
Attention to course of study	☐	☐	x	☐	☐	*Quite flexible—allows students to wander to different topics*
Knowledge of subject matter	☐	☐	☐	☐	x	*Does not know material—has to question pupils to gain knowledge*

E. PROFESSIONAL ATTITUDE

	1	2	3	4	5	Comments
Professional ethics	☐	☐	☐	☐	x	*Does not belong to professional association or PTA*
In-service training	☐	☐	☐	☐	x	*Complete failure here—has not even bothered to attend college*
Parent relationships	☐	☐	☐	☐	x	*Needs to improve in this area—parents are trying to get rid of him*

RECOMMENDATION: *Does not have a place in Education. Should not be rehired.*

Physics Teacher **13**, 86 Feb 1975

From *Physics Teacher* (November 1975). To save you a trip to the encyclopaedia, Newton was born in Woolthorpe on 25 December 1642.

Proclamation

WHEREAS: Isaac Newton is the founder of classical Physics, and,

WHEREAS: the birth of such an intellectual giant should be honoured,

BE IT Therefore Resolved that the Physics classes will be dismissed from school on the anniversary of his birth.

_____

Principal

Signature

This is a bit of trivia I sometimes use to make all students aware there is a physics department. I put this in the school paper with the signature (and approval) of the Principal. Perhaps it may be of some use to others.

MICHAEL SCOTT

Free thinking

ALEXANDER CALANDRA

From *The Saturday Review* (21 December 1968) p 60.

Some time ago I received a call from a colleague who asked if I would be the referee on the grading of an examination question. He was about to give a student a zero for his answer to a physics question, while the student claimed he should receive a perfect score and would if the system were not set up against the student. The instructor and the student agreed to an impartial arbiter, and I was selected.

I went to my colleague's office and read the examination question: 'Show how it is possible to determine the height of a tall building with the aid of a barometer.'

The student had answered: 'Take the barometer to the top of the building, attach a long rope to it, lower the barometer to the street, and then bring it up, measuring the length of rope. The length of the rope is the height of the building.'

I pointed out that the student really had a strong case for full credit, since he had answered the question completely and correctly. On the other hand, if full credit were given, it could well contribute to a high grade for the student in his physics course. A high grade is supposed to certify competence in physics, but the answer did not confirm this. I suggested that the student have another try at answering the question. I was not surprised that my colleague agreed, but I was surprised that the student did.

I gave the student six minutes to answer the question, with the warning that his answer should show some knowledge of physics. At the end of five minutes, he had not written anything. I asked if he wished to give up, but he said no. He had many answers to the problem; he was just thinking of the best one. I excused myself for interrupting him, and asked him to please go on. In the next minute he dashed off his answer which read:

'Take the barometer to the top of the building and lean over the edge of the roof. Drop the barometer, timing its fall with a stopwatch. Then, using the formula $S = \frac{1}{2}at^2$, calculate the height of the building.'

At this point, I asked my colleague if *he* would give up. He conceded, and I gave the student almost full credit.

On leaving my colleague's office, I recalled that the student had said he had other answers to the problem, so I asked him what they were. 'Oh, yes' said the student. 'There are many ways of getting the height of a tall building with the aid of a barometer. For example, you could take the barometer out on a sunny day and measure the height of the barometer, the length of its shadow, and the length of the shadow of the building, and by the use of simple proportion, determine the height of the building.'

'Fine' I said. 'And the others?'

'Yes' said the student. 'There is a very basic measurement method that you will like. In this method, you take the barometer and begin to walk up the stairs. As you climb the stairs, you mark off the length of the barometer along the wall. You then count the number of marks, and this will give the height of the building in barometer units. A very direct method.

'Of course, if you want a more sophisticated method, you can tie the barometer to the end of a string, swing it as a pendulum, and determine the value of 'g' at the street level and at the top of the building. From the difference between the two values of 'g', the height of the building can, in principle, be calculated.

'Finally,' he concluded 'there are many other ways of solving the problem. Probably the best' he said 'is to take the barometer to the basement and knock on the superintendent's door. When the superintendent answers, you speak to him as follows: "Mr Superintendent, here I have a fine barometer. If you will tell me the height of this building, I will give you this barometer."'

At this point, I asked the student if he really did not know the conventional answer to this question. He admitted that he did, but said that he was fed up with high school and college instructors trying to teach him how to think, to use the 'scientific method', and to explore the deep inner logic of the subject in a pedantic way, as is often done in the new mathematics, rather than teaching him the structure of the subject. With this in mind, he decided to revive scholasticism as an academic lark to challenge the Sputnik-panicked classrooms of America.

33

A revised PhD comprehensive

From Pennsylvania State University *Intercom* (17 November 1977) p 2.

Instructions: Read each question carefully. Answer *all* questions. Time limit: 4 hours. Begin immediately.

History. Describe the history of the papacy from its origins to the present day, concentrating especially but not exclusively on its social, political, economic, religious and philosophical impact on Europe, Asia, America and Africa. Be brief, concise and specific.

Medicine. You have been provided with a razor blade, a piece of gauze, and a bottle of Scotch. Remove your appendix. Do not suture until your work has been inspected. You have 15 minutes.

Public speaking. 2500 riot-crazed aborigines are storming the classroom. Calm them. You may use any ancient language except Latin and Greek.

Biology. Create life. Estimate the differences in subsequent human culture if this form of life had developed 500 million years earlier, with special attention to its probable effect on the English parliamentary system. Prove your thesis.

Music. Write a piano concerto. Orchestrate and perform it with flute and drum. You will find a piano under your seat.

Psychology. Based on your knowledge of their works, evaluate the emotional stability, degree of adjustment, and repressed frustrations of each of the following: Alexander of Aphrodisias, Ramses II, Gregory of Nicea, and Hammurabi. Support your evaluation with quotations from each man's work, making appropriate references. It is not necessary to translate.

Sociology. Estimate the sociological problems which might accompany the end of the world. Construct an experiment to test your theory.

Engineering. The disassembled parts of a high-powered rifle have been placed in a box on your desk. You will also find an instruction manual, printed in Swahili. In ten minutes a hungry Bengal tiger will be admitted to the room. Take whatever action you feel appropriate. Be prepared to justify your decisions.

Economics. Develop a realistic plan for refinancing the national debt. Trace the possible effects of your plan in the following areas: Cubism, the Donatist controversy, the wave theory of

34

light. Outline a method for preventing these effects. Criticize this method from all possible points of view. Point out the deficiencies in your point of view, as demonstrated in your answer to the last question.

Political science. There is a red telephone on the desk beside you. Start World War III. Report at length on its socio-political effects, if any.

Epistemology. Take a position for or against truth. Prove the validity of your position.

Physics. Explain the nature of matter. Include in your answer an evaluation of the impact of the development of mathematics on science.

Philosophy. Sketch the development of human thought; estimate its significance. Compare with the development of any other kind of thought.

General knowledge. Describe in detail. Be objective and specific.

Syllabus for a detective story written by a physics professor

From A E S Green *American Journal of Physics* **43** 222 (1975).

Chapter I	Origins of Law in Babylon
II	Constitution of United States
III	Basic Organization of Police Department
IV	Elements of Courtroom Practice
V	Theory of Fingerprints
⋮	
Chapter XXX	(Last Page) The Corpse (Solution Left to the Student)

Sales: *Zero*

H R CRANE

Challenging experiments

JOHN H REYNOLDS

From *Physics Today* **28** 64 (1975).

I recently pre-enrolled students in a sophomore course at Berkeley, Physics 4E. It is a day-long process wherein students come in to fill out cards and pick up 'handouts'. Among the things I handed out was a list of the laboratory experiments together with some comments about them. When I came to those that don't work very well as a rule, I fell back on the usual professorial gambit of describing them as 'more challenging'. I then grandly departed, leaving the rest of the chore to the graduate teaching assistants. Returning at 4:30 pm to close the pre-enrolment I found that my assistants had compiled their own list of 'more challenging labs' on the blackboard:

(1) Photograph vortex lines of ^{4}He.

(2) Measure neutrino flux from the Sun; compare with theory.

(3) Produce element 106; measure its halflife.

(4) Measure gravitational wave flux from Crab Pulsar.

(5) Build atomic bomb; test for ecological damage to environment.

(6) Construct a 1 gram black hole; compare radiation flux to Siberian meteor of 1908.

(7) Build a working fusion reactor.

(8) Measure Young's modulus of metallic hydrogen.

(9) Establish radio contact with extra-terrestrial life.

The authors of this list were Stephen Pollaine and Jerry Turney, perhaps with a little help from their friends.

Images

W S FRANKLIN AND B MCNUIT

From 'A calendar of leading experiments' *American Physics Teacher* **4** 131 (1936).

Why is an object seen erect when its image on the retina is inverted? In answer to this question the equally sensible question is sometimes asked: When one hears a baby cry with two ears, why does one not take it for twins?

Mission statements

RICHARD CHAIT

Adapted from
'Mission madness
strikes our colleges'
*Chronicle of Higher
Education* (16 July
1979).

Committees and commissions are at work everywhere on declarations of goals and statements of purpose for colleges and universities. Campus groups with different names are pondering the same questions: What is our mission? Who are our clients? Where are we headed? We are on the verge of mission madness. Mission statements are not so much valueless as overvalued. We make too much of the process and the product. Why?

First, governing boards are keen on mission statements. As the traditional guardians of the college's mission, trustees naturally worry about the state of the statement. Most board members can recite the college's mission for the next five years. Unfortunately, only a handful can suggest whether the goals set five years earlier have now been achieved. No matter; the college has an up-to-date mission statement.

Second, presidents are keen on mission statements. No one accepts a presidency to maintain the status quo. Presidents intend to be leaders, change agents, and what better way to initiate change than to re-examine the college's mission. Thus, almost as a ritualistic sequel to investiture, the new president convenes a committee to rethink the university's future.

Third, regional accrediting associations are keen on mission statements. To gain or maintain accreditation, a college needs a mission statement. No statement means no accreditation, and no accreditation means no future.

Trustees, presidents, and accreditors are likely to exaggerate the importance of mission statements; an enrolment problem guarantees excessive concern. As enrolments start to dwindle, an institutional identity crisis ensues. The campus community frets until suddenly everyone recognizes the obvious; we need a new mission statement. Inevitably, the notion of a new, clearer and singular statement gathers momentum and assumes a quality larger than life, almost as if its very appearance will solve all problems. When the document finally returns from the printer, everyone feels greatly relieved. A mission in print is a mission in fact. The crisis has passed.

Distinctiveness derives more from execution than from mission, more from what a college does and less from what it purports to be. In the final analysis, the mission will always be, in some sense, survival. Missions will be adapted to the market to ensure the organization's continuation. But I am less troubled by responsiveness to the marketplace than by the linguistic and philosophical gymnastics undertaken to merge the 'new thrust' with the old mission.

The best way to survive, even flourish, may be to worry a little less about mission statements and a little more about actions. Of course the two are related, but we have overworked the former, perhaps because it is not as easy to attend to the latter.

The rank order on campus

PEARL G ALDRICH

From *Chronicle of Higher Education* **19** 2 (14 January 1980).

An Arrogance of Deans
A Complaisance of Professors
An Ambition of Associate Professors
A Jitter of Assistant Professors
A Bewilderment of Instructors
A Hunger of Part-timers
A Starvation of Teaching Assistants

altogether,

A Paranoia of Faculty

Budget

From Isaac Asimov *Treasury of Humour* (London: Woburn Press) 1971.

UNIVERSITY PRESIDENT: 'Why is it that you physicists always require so much expensive equipment? Now the Department of Mathematics requires nothing but money for paper, pencils, and erasers . . . and the Department of Philosophy is better still. It doesn't even ask for erasers.'

The university hierarchy—who's on top?

ANON

The Dean
  Leaps tall buildings in a single bound
  Is more powerful than a locomotive
  Is faster than a speeding bullet
  Walks on water
  Gives policy to God

The Department Head
  Leaps short buildings in a single bound
  Is more powerful than a switch engine
  Is just as fast as a speeding bullet
  Walks on water if sea is calm
  Talks with God

Professor
  Leaps short buildings with a running start and favourable
  winds
  Is almost as powerful as a switch engine
  Is faster than a speeding BB
  Walks on water in an indoor swimming pool
  Talks with God if special request is approved

Associate Professor
  Barely clears a Quonset hut
  Loses tug of war with locomotive
  Can fire a speeding bullet
  Swims well
  Is occasionally addressed by God

Assistant Professor
  Makes high marks on the walls when trying to leap tall
  buildings
  Is run over by locomotives
  Can sometimes handle a gun without inflicting self-injury
  Talks to animals

39

Graduate Student
  Runs into buildings
  Recognizes locomotives two out of three times
  Is not issued ammunition
  Can stay afloat with a life jacket
  Talks to walls

Undergraduate Student
  Falls over doorstep when trying to enter building
  Says look at the choo-choo
  Wets himself with a water pistol
  Plays in mud puddles
  Mumbles to himself

Department Secretary
  Lifts buildings and walks under them
  Kicks locomotives off the tracks
  Catches speeding bullets in her teeth and eats them
  Freezes water with a single glance
  She is God.

View from the bottom

C E K MEES

The best person to decide what research work shall be done is
the man who is doing the research, and the next best person is
the head of the department, who knows all about the subject
and the work; after that you leave the field of the best people
and start on increasingly worse groups, the first of these being
the research director, who is probably wrong more than half
the time; and then a committee, which is wrong most of the
time; and finally a committee of vice presidents of the
company, which is wrong all the time.

[*Dr Mees later served as head of research at Eastman Kodak for several
decades.*]

On the problem of innovation

Condensed from
'After Vietnam:
an approach to
future wars of
national liberation'
*Southeast Asian
Perspective* **4** 1–29
(December 1971).

What was really amazing was the speed with which the Americans adapted themselves to modern warfare. They were assisted in this by their tremendous practical and material sense and by their lack of all understanding for tradition and useless theories.

FIELD MARSHALL ERWIN ROMMEL, quoted in *Rommel* by Desmond Young (New York: Berkley) 1964, p 159.

But since World War II, the doctrine developed by Americans fighting *that* war (while they were ripping up older doctrine) has become canonized. There is a need and use for doctrine, but not if it is graven on tablets of stone and treated as if it came from Mount Sinai. Doctrine should come in a looseleaf notebook.

Why is innovation resisted? My dictionary defines 'innovate' as follows: to bring in something new, or to make changes in anything established. Resistance arises more from the second usage than the first: 'making changes in anything established' is equivalent to rocking the boat. In practice, the maxim should read 'Never rock a sinking boat.'

Somewhat simplified, there are two kinds of people in the military services (their civilian counterparts are more numerous): those who are 'good at' expensive, massive technologically rich, complicated, but routine operations; and those who prefer to operate in small non-standard ways, who know how and why to improvise, who like elegance of operation.

An example illustrating what I mean by elegance comes from a memorable experience 25 years ago. A then senior colonel in reconaissance was placed in charge of the air photographic portion of Operation Crossroads, the 1946 atomic bomb test at Bikini. His previous experience did not at all equip him to cope with the peculiar photographic problems to be encountered in this kind of operation. Quite properly, and intelligently, he set out to see us at Wright Field; I was the project engineer for this exercise. He asked me to prepare a detailed plan, and gave me a week or so to do it. I came back with a proposal involving the use of only two C-54 aircraft, dozens of cameras and instruments in each, with each aircraft a back-up for the other. Assuming that I would be allowed to choose the people, I told him that I would assemble a crew of about thirty or so people from Wright Field, all of whom had experience in the various aspects of aerial photography, all of whom could maintain their own equipment, do the dark-room and technical work; and that in this group there would be a sufficient number to do the scientific analysis and write the required reports. I told him 'It will be an elegant operation, and will show how much can be done with a few people, if they're the right ones.' He looked at me for a moment before replying, 'Look, Katz, you don't understand. I don't want an elegant operation. I want a big outfit.' He got the big outfit, with thirteen aeroplanes and 850 people, but all the work was done by the two aircraft we had discussed.

It seems to be a property of organizations that no one, civilian or military, would prefer to have fewer people work for him tomorrow than he has working for him today. But quantity is hardly ever a substitute for quality. As long as promotion—whether more stars or higher civil service ratings—is thought to be dependent on increased numbers of people underneath, there will be a drive toward bigger organizations. Promotion, after all, is the sole way of rewarding progress or showing profit in government. We need a system where one is rewarded for 'going small'.

From *Nature* **218** p 797 (1966).

In the days of old and insects bold
(Before bats were invented)
No sonar cries disturbed the skies
Moths flew uninstrumented.

J D PYE

An undergrad from Trinity
Computed the square of infinity
But so great were the digits
That he got the figits
And turned from maths to divinity.

ANON

The Dreistein case

J LINCOLN PAINE

From *The Washington Star* (17 November 1957). J Lincoln Paine is the pen name of Arnold Kramish.

[*On 2 August 1939, Albert Einstein wrote to Franklin D Roosevelt a letter calling the President's attention to the implications and promise of recent researches in nuclear physics. The world-altering results are well known. But a pseudonymous scientist, who has had personal experience with the labyrinthine ways of the Government, suggests that such a letter might not invariably meet with such a response.*]

Advanced Research Institute
Cambridge
Massachusetts

2 August 1961

The President of the United States
The White House
Washington DC

Esteemed Sir:

Some recent work by my colleague, Professor Hauck of Pretoria, has been communicated to me in manuscript. His findings lead me to believe that scientists may be able to counteract the forces of gravity in the near future. Undoubtedly, if Hauck's new discoveries are further developed and applied, a vast new area of space exploration and missile development will open.

The situation which has arisen seems to call for watchfulness and, if necessary, quick action on the part of the Administration. My colleagues here have urged me to bring this obviously significant development to the attention of the appropriate government authorities. I believe, therefore, that it is my duty to bring to your attention some of the scientific facts which are attached in a separate memorandum.

Of course, my colleagues and I offer our full services towards the further development of this discovery.

Very truly yours,

Egbert Dreistein

THE WHITE HOUSE
Office of the Special Assistant to the President

16 August 1961

To: The Secretary of Defense

Attached is copy of letter from Prof. Egbert Dreistein. Draft reply for my signature. Be polite. Incidentally, is there anything to this?

Grant Quincey

INTER-OFFICE MEMORANDUM

To: Colonel T Lee, OPS   From: The Secretary
Date: 2 September 1961   Ref: CPT-201/1

Prepare reply to attachment. Is the institute under contract to the DOD? Quote me their budget figures for the last three fiscal years.

Official Use Only
INTER-OFFICE MEMORANDUM

To: The Secretary   From: Colonel T Lee, OPS
Date: 29 June 1962   Ref: CPT-201/179

The matter referred to in your memorandum CPT-201/1 of 2 September 1961 has been referred to an Inter-service Ad Hoc Committee of staff-rank representatives. The committee concurred that there was no consensus on the problem.
    Individual views were as follows:

(I) The Army feels that ordinary gravity is not fully understood yet and sees little purpose in extending studies into the field of anti-gravity.

(II) The Air Force has been conducting small-scale research on anti-gravity at the TOP SECRET level. However, since it is impossible to extend the concept to fit existing weapons systems, a low priority has been assigned.

(III) The Navy has recommended a high priority to anti-anti-gravity investigations under the code name of PLOP.

There is no record in DOD files of a facility clearance for the Advanced Research Institute. Prof. Dreistein has never applied for a 'Q' clearance. Given the sensitive nature of the anti-gravity question and the extenuating circumstances, the attached draft reply to Prof. Dreistein has been made as clear as classification permits.

The committee reached agreement on a single point: Prof. Dreistein should not be encouraged. A permanent subcommittee has been set up to provide similar assistance in expediting the handling of any future suggestions from members of the scientific community.

DEPARTMENT OF DEFENSE
Office of the Secretary

2 July 1962

To: Special Assistant to the President

In reference to your request of 16 August 1961, attached is draft reply to Prof. Egbert Dreistein.

The receipt of Prof. Dreistein's letter has stimulated reexamination of the status of anti-gravity research in the Department of Defense. Estimated future budgetary allocations for that type of research do not warrant continuation of the projects which have been under way. Accordingly, I have issued an order that they be curtailed.

Frank Watt

THE WHITE HOUSE
Office of the Special Assistant to the President

5 July 1962

Prof. Egbert Dreistein
The Advanced Research Institute
Cambridge, Massachusetts

Dear Professor Dreistein:

The President has directed me to reply to your letter of 2 August 1961. We thank you for your interest and assure you that the matter has been investigated by appropriate government agencies.

Your patriotic interest is very much appreciated and the President is always interested in receiving stimulating ideas of that nature.

Yours very truly,

Grant Quincey

Moscow, 5 August (1964).—A Soviet spokesman announced today that a manned space station has been established as a satellite around Mars and is now observing landing conditions on that planet.

The achievement was credited to the revolutionary discoveries of Professor Otto Hauck, formerly of South Africa and now in the Soviet Union. He has been awarded three Lenin prizes for his work . . .

THE WHITE HOUSE

6 August 1964

Prof. Egbert Dreistein
The Advanced Research Institute
Cambridge, Massachusetts

Dear Professor Dreistein:

My advisers report to me that you have been interested in the subject of anti-gravity research. Because of the grave circumstances in which our Government finds itself as a result of the announcement from Moscow yesterday, I am asking you to lead a new high-priority project in that direction.

If you will come to Washington in the early part of next week, a briefing will be arranged by representatives of the military services and the Central Intelligence Agency who will be able to give you a little of the historical development of Professor Hauck's work.

I, as President, personally hope that you and your colleagues will rise to the challenge of this new emergency.

Yours very truly,

Horatio Calvin

Science is the best way of satisfying the curiosity of individuals at government expense.

L A AKTSIMOVICH *Novy Mir* 1 (1967)

intoxication must have been the common substance: namely *the soda water!*'

Appleton was pleased but a little surprised at the huge success of the story. Only later did he learn that the Crown Prince drank only soda water—'one of those unexpected bonuses which even the undeserving get from Providence from time to time,' as he put it.

Hypothesis

A S PARKES

From *Perspectives in Biology and Medicine* **7** 366 (1958). It has often been pointed out that a very good example of the use of hypothesis is provided by Columbus' discovery of America; it has many features of a classic discovery in science. Columbus was obsessed with the idea that if the Earth were round he could reach the East Indies by sailing west.

Notice the following points:

(a) The idea was by no means original, but he had obtained some additional scraps of information.

(b) He met great difficulties in getting someone to provide the money as well as making the actual experiment.

(c) He did not find the expected new route but, instead, found a new half of the World.

(d) Despite all evidence to the contrary, he clung to the belief that he had found a new route to the Orient.

(e) He got little credit or reward during his lifetime.

(f) Evidence has since been brought forward that he was by no means the first European to reach America.

I await your sentence with less fear than you pass it. The time will come when all will see what I see.

GIORDANO BRUNO

Finagle's laws

. . . or why nothing in Research & Development happens the way it should

From *IRE Student Quarterly* (September 1958) p 42. Collected by John W Campbell, Editor, *Astounding Science Fiction* and his readers.

FIRST LAW

If anything can go wrong with an experiment, it will.

SECOND LAW

No matter what result is anticipated, there is always someone willing to fake it.

THIRD LAW

No matter what the result, there is always someone eager to misinterpret it.

FOURTH LAW

No matter what occurs, there is always someone who believes it happened according to his pet theory.

THE LAW OF THE TOO SOLID GOOF

In any collection of data, the figure that is most obviously correct—beyond all need of checking—is the mistake.

Corollary I

No one whom you ask for help will see it either.

Corollary II

Everyone who stops by with unsought advice will see it immediately.

ADVICE TO EXPERIMENTERS

(1) Experiments must be reproducible—they should all fail in the same way.

(2) First draw your curves—then plot the readings.

(3) Experience is directly proportional to equipment ruined.

(4) A record of data is useful—it indicates you've been working.

(5) To study a subject best, understand it thoroughly before you start.

(6) In case of doubt, make it sound convincing.

(7) Do not believe in miracles—rely on them.

(8) Always leave room to add an explanation when it doesn't work. (This open-door policy is also known as the Rule of the Way Out.)

A mathematical notation of Finagle's work has also been developed. Here, however, there seems to be some confusion, because two other names enter the picture: 'fudge' and 'diddle' factors are also used to considerable advantage by scientists and engineers.

However, John W Campbell feels there is a different basic structure behind the Finagle, fudge, and diddle factors. The Finagle factor, he claims, is characterized by changing the Universe to fit an equation. The fudge factor, on the other hand, changes the equation to fit the Universe. And finally, the diddle factor changes things so that the equation and the Universe appear to fit, without making any real change in either.

For example, the planet Uranus was introduced to the Universe when Newtonian laws couldn't be made to match known planetary motions. This is a beautiful example of the application of the Finagle factor.

Einstein's work leading to relativity was strongly influenced by the observed facts about the orbit of Mercury. Obviously a fudge factor was introduced.

The photographer's use of a 'soft-focus' lens when taking portraits of women over 35 is an example of the diddle factor. By blurring the results, photographs are made to appear to match the facts in a far more satisfactory manner.

To our knowledge, this is the first clear enunciation of the scientific method. All our vast sum of human knowledge has been derived with these as the basic tools. By having them in writing for the first time, perhaps our children can build even better futures than the best we envision today.

In all matters of opinion, our adversaries are insane.

MARK TWAIN

An illustration of Murphy's Law applied to surface science: After a vacuum flange has been secured by 16 bolts, it will be discovered that the sealing gasket has been left out (J A Venables).

Descent of man

D L AUSTIN AND J D BU'LOCK

From *Science* **163** 623 (1969).

The truly fascinating article by Kellog ('Communication and language in the home-raised chimpanzee' 25 October 1969, p 423) prompts us to record remarkably similar gestural communications in laboratory-raised *Homo sapiens* (var. *postgraduatensis*) observed over a period of years in this and similar institutions, and tabulated in summary form below.

Gesture signals of experimenter.

Behaviour pattern	Interpretation
Biting or chewing at clothing, or ball points	It worked last time
Concealing cup	Here comes the boss
Climbing into high chair	I am the boss
Removing white coat	I quit
Taking hand of (visiting) experimenter and hanging on to it	I need a job
Putting hand to bottle	What the hell
Throwing self to floor	Reaction is highly exothermic

Science develops from the play of a civilized society.

ERWIN SCHRÖDINGER

It is much easier to make measurements than to know exactly what you are measuring.

J W N SULLIVAN, 1928

It is a capital mistake to theorize before one has data. Insensibly one begins to twist facts to suit theories, instead of theories to suit facts.

ARTHUR CONAN DOYLE

Literacy in an age of technology

GEORGE STEINER

Condensed from 'A
future literacy'
Atlantic Monthly
(August 1971).
Changes of idiom between generations are a normal part of
social history. Previously, however, such changes and the
verbal provocations of young against old have been variants
on an evolutionary continuum. What is occurring now is new;
it is an attempt at a total break. Deliberate violence is being
done to those primary ties of identity and social cohesion
produced by a common language.

But are there no other literacies conceivable, 'literacies' not
of the letter? Popular musics have their semantics, their theory
of genres, their intricate play-offs of esoteric against canonic
types. In short, the vocabularies, the contextual behaviour
patterns of pop and rock, constitute a genuine *lingua franca*, a
'universal dialect' of youth. Everywhere a sound-culture
seems to be driving back the old authority of verbal order.

If music is one of the principal 'languages outside the word',
mathematics is another. Any argument on postclassic culture
and on future literacy will have to address itself, decisively, to
the role of the mathematical and natural sciences. More than
90% of all scientists known to human record are now living.
Between now and 1990, according to a recent projection, the
number of monographs published in mathematics, physics,
chemistry and biology will, if aligned on an imaginary shelf,
stretch to the moon.

Less tangibly, but more significantly, it has been estimated
that some 75% of the most talented individuals in the
developed nations, of the men and women whose measurable
intelligence comes near the top of the curve in the community,
now work in the sciences. Politics and the humanities thus seem
to draw on a quarter of the optimal mental resources in our
societies, and recruit largely from below the line of excellence.

One can identify [several] areas of maximal pressure,
points at which pure science and technological realization will
alter basic structures in both private and social life. The
revolutions of awareness that will result from full-scale com-
puterization and electronic data processing can only crudely
be guessed at. In advanced societies, the electronic data bank
is becoming the pivot of military, economic, sociological and
archival procedures. Another main area is that of large-scale
ecological modification. Control of weather, locally at least, is

now conceivable, as is the economic exploitation of the continental shelves and of the deeper parts of the sea.

To be ignorant of these scientific and technological phenomena, to be indifferent to their effects on our mental and physical experience, *is to opt out of reason*. A view of postclassic civilization must, increasingly, imply a vision of the sciences, of the language-worlds of mathematical and symbolic notation. Today our dialectics are binary.

To have some personal *rapport* with the sciences is, very probably, to be in contact with that which has the most force of life and comeliness in our reduced condition. Touch on even its more abstruse regions and a deep elegance, a quickness and merriment of the spirit come through. A 'poetry of facts' and realization of the miraculous delicacies of perception in contemporary science already inform literature at those nerve-points where it is both disciplined and under the stress of the future.

Modern science is centrally mathematical. Even a modest mathematical culture will allow some approach to what is going on. The notion that one can exercise a rational literacy in the latter part of the twentieth century without a knowledge of calculus, without some preliminary access to topology or algebraic analysis, will soon seem a bizarre archaicism. These styles and speech-forms from the grammar of number are already indispensable to many branches of modern logic, philosophy, linguistics and psychology. They are the language of feeling where it is, today, most adventurous. A new hierarchy of menial service and stunted opportunity may develop among those whose resources continue to be purely verbal. There may be 'word helots'.

The history of science, moreover, permits of a less demanding access, yet one that leads to the centre. A modest mathematical culture is almost sufficient to enable one to follow the development of celestial mechanics and of the theory of motion until Newton and Laplace. It takes no more than reasonable effort to understand, at least along major lines, the scruple, the elegance of hypothesis and experiment which characterize the modulations of the concept of entropy from Carnot to Helmholtz. The genesis of Darwinism and the subsequent re-examinations which lead from orthodox evolutionary doctrine to modern molecular biology are one of the 'very rich hours' of the human intellect. Yet much of the material, and many of its philosophical implications, are accessible to the layman. This is so, to a lesser degree, of some part of the debate between Einstein, Bohr, Wolfgang Pauli and Max Born—from each of whom we have letters of matchless honesty and personal commitment—on the issue of anarchic indeterminacy of subjective interference in quantum physics. Here are topics as crowded with felt life as any in the humanities. The absence of the history of science and technology from the school syllabus is a scandal.

Overwhelmingly, today, science is a collective enterprise in which the talent of the individual is a function of the group. But, as we have seen, more and more of current radical art and anti-art aspire to the same plurality. The deep divergence between the humanistic and scientific sensibilities is one of temporality. Very nearly by definition, the scientist knows that tomorrow will be in advance of today. A twentieth-century schoolboy can manipulate mathematical and experimental concepts inaccessible to a Galileo or a Gauss. For a scientist the curve of time is positive. Inevitably, the humanist looks back. A natural bent of feeling will lead him to believe, perhaps silently, that achievements of the past are more radiant than those of his own age. The proposition that 'Shakespeare is the greatest, most complete writer' mankind will ever produce is a logical and almost a grammatical outrage. But it carries conviction. There is a profound logic of sequent energy in the arts, but not an additive progress in the sense of the sciences. No errors are corrected or theorems disproved. Because it carries the past within it, language, unlike mathematics, draws backward. This is the meaning of Eurydice. Because the realness of his inward being lies at his back, the man of words, the singer, will turn to the place of necessary shadows. For the scientist, time and light lie before.

Here, if anywhere, lies the division of the 'two cultures' or, rather, of the two orientations. Anyone who has lived among scientists will know how intensely this polarity influences life-style. Their evenings point to tomorrow, *e santo è l'avvenir.*

57

How to live with the Philistines

HENRY E DUCKWORTH

Condensed from *Physics Today* **21** 54 (1968). Based on a speech delivered at a joint meeting of the Sociedad Mexicana de Fisica, the American Physical Society and the Canadian Association of Physicists, held in Toronto, 1967.

[*The physicist is surrounded by outsiders, be they relatives, other scientists, humanities professors or officials. He must increase his interaction with them to survive.*]

To clarify my topic I quote the *Concise Oxford Dictionary* definition of the word 'Philistine': 'one of an alien warlike people in South Palestine who harassed the Israelites; . . . an outsider; (or in general) an uncultured person, one whose interests are material and commonplace.' I shall use the word in the sense of an outsider. Thus, a rewording of the subject for purposes of clarification would be 'How to live with non-physicists.' I shall have space to refer to only a few of the principal types.

THE WIFE

This case is non-trivial and its crucial interaction is during the premarital or courtship period. If a satisfactory marriage is to ensue, it is imperative at this time that the customary frivolities such as dances and movies be avoided. Instead the lucky girl should be taken to the laboratory, where she may pass many a pleasant evening watching her friend with admiration as he repairs apparatus or makes readings. A union based on such a courtship is bound to be agreeable to the physicist, provided he takes the additional precaution of including in his marriage proposal the immortal lines of Richard Lovelace (1618–1658):

Tell me not, sweet, I am unkind,
That from the nunnery
Of thy chaste breast and quiet mind
To pump and 'scope I fly.

True, a new mistress now I chase,
My own on-line computer;
And with a stronger faith embrace
My solid-state transducer.

Yet this inconstancy is such
As you too shall adore;
I could not love thee, dear, so much,
Loved I not physics more.

58

I feel that there is no basic difficulty in living with this type of non-physicist, provided one starts out on the right foot.

OTHER SCIENTISTS

A second type of non-physicist with whom arrangements for co-existence should be fostered is the professor or research worker in a field other than physics. As sciences such as biology, chemistry, metallurgy and geology become more sophisticated and approach their problems in an increasingly fundamental way, they draw increasingly on the concepts, techniques and instrumentation of physics. Many of the problems faced by non-physicists could benefit from the active participation of physicists themselves. Thus it was a physicist who had specialized in the study of electric currents and a biologist who had specialized in the study of the heart who combined to discover the law for the circulation of the blood, namely that it flows down one leg and up the other.

I have been impressed by the relevance of the problems the biologists have been tackling and by their resolute approach to them. Thus at the University of Toronto biologists are attempting to solve directly the problems arising from the population explosion by investigating methods of making birth control retroactive. In the interest of science another intrepid investigator lived for 28 days on dehydrated food. Unfortunately at the end of the period he was caught in a rainstorm and gained 108 pounds in ten minutes. I do not think that we as a group, for whatever reason, have taken an initiative in interdisciplinary research programmes commensurate with our potential contribution to them. In other words, we have not fully utilized this important device for living with the Philistines.

HUMANISTS

There are, of course, other scholarly Philistines with whom we come in contact, particularly in university communities, such as professors of English, philosophy, history, and so on. In the eyes of many of these we appear as barbarians, powerful barbarians it is true, but the more barbaric on that account. An intellectual chasm is growing between physicists and these humane or semihumane non-physicists. We were told a decade ago that the cure was a simple 50–50 arrangement: humanize the scientists and simonize the humanists. I feel now we must go much more than half way. There is a resentful suspicion that science is taking over, that it receives the lion's share of the pie and that the others are being forced to the wall. I do not have the space here to suggest methods by which we might stem this tide, but the whole concept of a university is at issue. A group of specialized institutes placed together in order to share common toilet facilities does not constitute a university. I urge each of you to combat this divisive trend by re-establishing diplomatic relations with this particular tribe of Philistines and then by searching assiduously for some common ground.

OFFICIALDOM

A defence counsel in a local police court once asked that his client be excused from attendance because, 'in the first place, he is a man of not very bright intellect. Secondly, he is employed on important government work.' In a somewhat exaggerated way, that remark expresses the view that many scientists have of many government officials. What is the other side of the coin? It may be typified by a description recently given by a government supervisor of one of his staff: 'This man is keenly analytic and his highly developed mentality could best be utilized in the research and development field. He lacks common sense.' This estrangement exists between the specialist who is absorbed in his own subject and the generalist who must reconcile the interests of many different constituencies, including political ones. In financial matters it is usually the generalist who has the upper hand. For that reason if for no other, it is important that we effect a reconciliation (or perhaps a conciliation, because I do not think we have had one before). To this end, we must study closely, for our own edification and for the edification of government officials, the role that physics does and should play in society. Then we must use whatever means we can devise (including petitions, cocktail parties, official delegations, fishing trips, newspaper interviews, seduction) to provide the generalist with an understandable picture of the contribution that physics actually makes. I realize that this is very easy to say and very hard to do, but its importance cannot be exaggerated. We must learn to live with these Philistines in order to survive, and we must do so by using their own idiom to transmit views that are conspicuous for common sense and complete defensibility.

A Philistine asks for equal time

SISTER JOHNELL DILLON

Condensed from *Physics Today* 22 39 (1969).

I feel an urge to speak on the side of the Philistines. As an English teacher who roomed for two years with a physics teacher, I feel that I am a qualified spokesman.

Although I am one of those people who increases the amount of entropy in a scientist's world through my ideas of the semantics of organization and unifying principles, still I have devoted myself humanly and selflessly, as best I can, to the pursuit of scientific learning.

I have groped through pitch-black labs at ungodly hours, squinted through slits and wiggled prisms until they were just right. My feeling of accomplishment when I really saw that spectrum—all colours, not just some—paralleled, if not exceeded, any joy of Marie Curie.

IN SOME RESPECTS, PRAGMATIC

I've supplied the round base of my lamp, ink bottle, vase and cup to aid in the construction of diagrams no compass ever knew. And once, helping with the *Sunday Times* crossword puzzle, I deftly and correctly offered the word 'ohm'. 'What's that?' said my mother. My scientific, humanistic response: 'You know, "What is ohm without a mother?"' It wasn't the answer of an empiricist, but it satisfied her for what it was worth—she was a humanist too.

I'd never suggest that I am scientific in any way, but I am pragmatic in some respects. Now, in one physics textbook, for example, there is a problem involving a station master who dropped an object on a platform as a train passed at 98% of the speed of light. The question asked for (to my mind irrelevant) complicated data as to the engineer's reaction to it all. I'm not absolutely sure of the exact speed of light, but I know this much: if that train was going *that* fast (and I doubt it), then the engineer scarcely saw the platform, let alone the station master *or* the object. But alas, my friend the physicist spent hours and pages on the solution of that improbable situation. What would a humanist have done?

WE WIRE THE LAMPS

My friend, unlike me, can discuss electromagnetism and Faraday's ice pail; she can readily distinguish AC from DC, anode from cathode and step-up from step-down transformer; deep in my heart, however, I believe she's afraid of electricity. So I wire the lamps, make the extensions, change the fuses— yes, and even sometimes plug in her experiments. When it comes to the mundane, it's a Philistine to the rescue every time!

Co-existing with a physicist, though, has been a unique learning experience, and I am eager to learn. My fund of knowledge has been enriched and expanded with items that I may never use, even in the most erudite society. I have a familiarity and even fondness for the Greek alphabet, the metric system and that piece of cat's fur. I listen and I care about cohesive forces and rarefactions. I try to correlate and integrate and I ask questions. I want to know why television tubes are the medium for some messages, how the second law of thermodynamics applies to communication and language, what makes the sound structure of language and when can molecular structure be analogous to literary structure. So I read physics magazines—almost in self defence—but I must admit I still have gaps, and I need to know. I ask.

No Philistine will ever penetrate your ranks. Not if you can help it, I think; not if the physicists I know are any indication of the others. My questions most frequently provoke the answer, 'I don't know,' which I immediately interpret to mean, 'I *do* know, but you'll never understand.' Now is that any way?

62

PHYSICISTS NEED PHILISTINES

If we Philistines fall by the wayside, you'll be sorry. Who'll cook and clean and mend and sew while you differentiate and integrate and calculate? Who'll correct your spelling while you adjust your sines and cosines and tangents? Who'll find your books in the library while you look for n? And who will, every now and then, slip a book of poems midst all your apparatus so that you may stabilize your equilibrium 'twixt heart and head? What *would* you do if we weren't there— Philistines, but none the less faithful?

Brain drain

JOAN BUTT

From *NPL News* 11 (21 November 1967).

A White Paper on our scientists
Is published to explain
What's made them leave old England's shores
When they've worked so hard to train.

Some say they wish to travel
Experience to gain;
Others thought of opportunity,
Position, wealth and fame.

So they left behind their favourite pubs,
Their English country and lane,
Their cricket, beer, and fish and chips
And constant English rain.

But did it need a Working Party
To make these facts so plain?
More cash, more scope, more everything
And in exchange—their brain.

But alas, I'm not a scientist,
And it's driving me insane;
There's all that money over there
And I've no brain to drain!

On Robert Boyle's tombstone in Dublin are incised the memorial words: 'Father of Chemistry and Uncle of the Earl of Cork.'

When hydrogen played oxygen,
And the game had just begun,
Hydrogen racked up two fast points,
But oxygen had none.

Then oxygen scored a single goal,
And thus it did remain.
Hydrogen 2 and oxygen 1,
Called off because of rain.

So much for chemistry . . .   ANON

Chemistry has been termed by the physicist as the messy part of physics, but that is no reason why the physicists should be permitted to make a mess of chemistry when they invade it.

FREDERICK SODDY

Quips from *Orbit*, Journal of the Rutherford High Energy Laboratory, Didcot, England.

For Sale: Do-it-yourself Nobel Prize kit—Accelerator (post-war model). Unused. Ideal for Enthusiast. Owner cannot afford upkeep. £11 000 000 o.n.o.

I would like so much to be in Oxford at that time, but I have not got any invitation . . .. My last hope is you. Do you think it is possible for you to seduce the Secretary of the Conference?—*A letter from Italy*

64

Pan's pipes

ROBERT LOUIS STEVENSON

From 'Virginibus Puerisque' *The Works of Robert Louis Stevenson* vol. 2 (New York: Charles Scribner's Sons) 1925, pp 140–3.

There are moments when the mind refuses to be satisfied with evolution, and demands a ruddier presentation of the sum of man's experience. At least, there will always be hours when we refuse to be put off by the feint of explanation, nicknamed science; and demand instead some palpitating image of our estate, that shall represent the troubled and uncertain element in which we dwell, and satisfy reason by the means of art. Science writes of the world as if with the cold finger of a starfish; it is all true; but what is it when compared to the reality of which it discourses, where hearts beat high in April, and death strikes, and hills totter in the earthquake, and there is a glamour over all the objects of sight, and a thrill in all noises for the ear, and Romance herself has made her dwelling among men? So we come back to the old myth, and hear the goat-footed piper making the music which is itself the charm and terror of things; and when a glen invites our visiting footsteps, fancy that Pan leads us thither with a gracious tremulo; or when our hearts quail at the thunder of the cataract, tell ourselves that he has stamped his hoof in the nigh thicket.

Seventeenth-century recognition of resonance

From *Flecknoe, an English Priest at Rome* by Andrew Marvell (1621–78).

Now as two instruments, to the same key
Being tuned by art, if the one touchèd be
The other opposite soon replies,
Moved by the air and hidden sympathies;
So while he with his gouty fingers crawls
Over the lute, his murm'ring belly calls,
Whose hungry guts to the same straitness twined
In echo to the trembling strings repined.

I cannot imagine how the clockwork of the universe can exist without a clockmaker.

VOLTAIRE

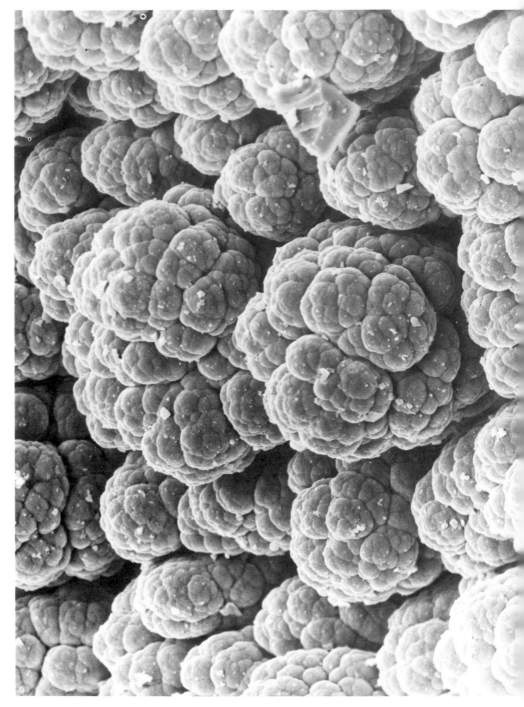

Demon theory of friction

ERIC M ROGERS

From *Physics for the Inquiring Mind* (Princeton: Princeton University Press) 1960.

How do you know that it is friction that brings a rolling ball to a stop and not demons? Suppose you argue for friction while a neighbour, Faustus, argues for demons. The discussion might run thus:

You: I don't believe in demons.

Faustus: I do.

You: Anyway, I don't see how demons can make friction.

Faustus: They just stand in front of things and push to stop them from moving.

You: I can't see any demons even on the roughest table.

Faustus: They are too small, also transparent.

Y: But there is more friction on rough surfaces.

F: More demons.

Y: Oil helps.

F: Oil drowns demons.

Y: If I polish the table, there is less friction and the ball rolls farther.

F: You are wiping the demons off; there are fewer to push.

Y: A heavier ball experiences more friction.

F: More demons push it; and it crushes their bones more.

Y: If I put a rough brick on the table I can push against friction with more and more force, up to a limit, and the block stays still, with friction just balancing my push.

F: Of course, the demons push just hard enough to stop you moving the brick; but there is a limit to their strength beyond which they collapse.

Y: But when I push hard enough and get the brick moving there is friction that drags the brick as it moves along.

F: Yes, once they have collapsed the demons are crushed by the brick. It is their crackling bones that oppose the sliding.[1]

[1] If Faustus has the equipment he should offer you a microphone attached to a glass table, with connections to an amplifier and loudspeaker. Then if you roll a steel ball along the table you will indeed hear noises like crushing demons.

◀ Not a bowl of raspberries, nor a cauliflower, but a photograph which shows gaseous silicon compounds doped with impurities precipitated as a glassy deposition on a quartz surface (Siemens Press Photo).

67

Y: I cannot feel them.

F: Rub your finger along the table.

Y: Friction follows definite laws. For example, experiment shows that a brick sliding along the table is dragged by friction with a force independent of velocity.

F: Of course, same number of demons to crush, however fast you run over them.

Y: If I slide a brick along the table again and again, the friction is the same each time. Demons would be crushed in the first trip.

F: Yes, but they multiply incredibly fast.

Y: There are other laws of friction: for example, the drag is proportional to the pressure holding the surfaces together.

F: The demons live in the pores of the surface: more pressure makes more of them rush out to push and be crushed. Demons act in just the right way to push and drag with the forces you find in your experiments.

By this time, Faustus' game is clear. Whatever properties you ascribe to friction he will claim, in some form, for demons. At first his demons appear arbitrary and unreliable; but when you produce regular laws of friction he produces a regular sociology of demons. At that point there is a deadlock, with demons and friction serving as alternative names for a set of properties—and each debater is back to his first remark.

You realize that friction has only served you as a name: it has established no link with other properties of matter. Then, as a modern scientist, you start speculating on the molecular or atomic cause of friction, and experimenting to test your ideas. Solids are strong; they hang together. Their component atoms must attract with large forces at short distances. When solid surfaces slide or roll on each other, small humps on one get within the range of atomic attractions of local humps on the other and they drag each other when the motion tries to separate them. Friction, then, may be an atomic dragging, which is likely to make one surface drag small pieces off the other. That has been investigated experimentally. After a copper block has been dragged along a smooth steel table, microphotographs show tiny copper whiskers torn off on to the steel. Also chemical tests show that a little of each metal rubs off on to the other.[2]

At last you have a good case for *friction*: it is a scientific name for some well-ordered behaviour that we can now link with other knowledge. It is atomic or molecular dragging, caused by the same forces that make wires strong and raindrops round. Its mechanism can be demonstrated by photographs and by chemical analysis. Its laws can even be predicted by applying our knowledge of elasticity to the small irregularities of surfaces. Friction has joined other phenomena in a general explanation.

And now we can state the full case against demons: they are arbitrary, unreasonable, multitudinous, and over-dressed. We need a special demon with peculiar behaviour to explain each natural event in turn: therefore we need many kinds and vast numbers of them. And we have to clothe them with special behaviours to fit all the facts. We now prefer something more economical and comfortable: a consistent body of knowledge, with strong ties to experiment—and with cross checks and interlinkages to assure us of validity- -all expressed in as few general laws as possible. Even where we meet new events that we cannot explain, we would rather speculate cautiously than invent a demon to calm some fear of mystery.

[2] We can even show that when a *copper* block rubs on another block of *copper*, tiny pieces of copper—invisibly small—are exchanged between the two blocks. No chemical analysis could tell one block's copper from the other's; so this interchange seems impossible to detect. Yet it is now easy. One can employ radioactivated tracer elements to follow the transfer.

Natural-science viva, c 1890

From *Oxford Outside the Guidebooks* by Falconer Madan, quoted in *The Oxford Book of Oxford* ed. Jan Morris (London: Oxford University Press) 1978.

EXAMINER: What is Electricity?

CANDIDATE: Oh, Sir, I'm sure I have learnt what it is—I'm sure I *did* know—but I've forgotten.

EXAMINER: How very unfortunate. Only two persons have ever known what Electricity is, the Author of Nature and yourself. Now one of the two has forgotten.

Physics instructor

JOHN N SHIVE

To the air 'I am the very model of a modern Major General' from *The Pirates of Penzance*.

[*Written in response to bellyaching from a class taught at Bell Telephone Laboratories around 1958.*]

I am the very model of a stuffy modern physicist.
To all the OETers I am Beelzebub's inquisicist.
The homework I assign is full of integrals elliptical,
With outside reading added via references cryptical.
Examination questions I propound with great facility
And point with blunt allusions to the quartile probability.
My students curse behind my back; the courses they are failing in
Are just the very ones I think they all should find clear sailing in.
And I am moved to risk predicting, when the final grades I think of,
The dismal end the whole Bell System seems to be upon the brink of!

My unapproachability and obvious pomposity
Are simply consequences of my stellar virtuosity.
The learned journals bristle with my articles oracular,
Recorded in the most incomprehensible vernacular.
The idioms of every tongue, from Greek to Scandinavian
Are just as much my dish as those of Pushtu or Mojavian.
Enthroned behind my desk piled high with Zeitschrifts full of verbiage,
In every language known to man, to none I yield subserviage.
My students haven't got a chance, and to prevent their penetration
I absolutely never risk a clear and simple explanation.

De Broglie waves and overlapping bands present no fear to me.
The theory of quantum states is absolutely clear to me.
With power series I'm at home; I might discourse a month or more
On integrating all the terms beyond the $(p + 1)$th or more.
An n-mesh problem I can solve with matrices symmetrical
Expanded into series which are hypergeometrical.
To me it is no task at all to hold within my cranium
The most elusive detail of the structure of germanium.
And how I love to cloud the story with evasive obfuscation
When in the classroom I unfold a messy lecture demonstration!

70

With penetrating insight I resolve the deepest mystery,
No matter whether it is in dynamics or transistory.
Performing a Laplace transform for me's an ordinary feat,
While Bessel and Legendre functions make the picture more complete.
My students indicate no pain about which I'm contented more
Than when they say, than what they knew, the latest quiz presented more.
Each lecture subject I attempt to hold, with laboured surety,
Secure and unrevealed behind the ramparts of obscurity.
With all these efforts I expend, the chances are at least good
To keep the Iv'ry Tower for the engineering priesthood.

'I can't stand it when he moves.'

Influential textbooks: *The Feynman lectures on physics*

From *Physics Today*
(November 1981)
p 758.

[*For the issue of* Physics Today *celebrating the 50th anniversary of The American Physical Society, the Editors invited Nobel Prizewinning physicists to comment on textbooks, or teachers, which especially influenced their early careers. Ivar Giaever's contribution follows.*]

'Should I write about Feynman?' I asked my friend. 'What for?' he said, 'To make Feynman famous?' It is difficult to say what for except that I really love the three big red books.

I had the fortune (misfortune?) of being exposed to physics for the first time when I was 29 years old. I found it frustrating because I could not find any books I liked. To my complaining, my friend said 'What did you expect, a graduate physics book written for Norwegian engineers?' I mumbled something intelligent. 'In the States' he continued 'we learn physics many times. Because it is regurgitated so often, in graduate school we concentrate on special points, and sometimes rather small ones at that.' Not much comfort to someone who did not know Planck's constant—not only the value but also the concept—and who marvelled that such famous physicists as Werner Heisenberg were still alive (I think).

Unfortunately, *The Feynman Lectures* were not published soon enough to help me out of my misery. However, when they appeared I slept with them under my pillow. (For this purpose, the paperback version is recommended.) I confess I have still not read every word in the books (probably neither has Feynman), but I have many favourite places. Take, for example, Feynman's delightful description of electrical power technology: 'Out of Boulder Dam come a few dozen rods of copper perhaps the thickness of your wrists that go for hundreds of miles in all directions. Small rods of copper carrying the power of a giant river . . ..' For someone interested in applied physics, it did not seem the average small point! Or, approximately in the middle of the second volume, he has collected *all* the equations of classical physics on *half* a page. How pleasurable to have all that knowledge distilled into a few equations, to know and understand them or at least have them look vaguely familiar. A little later in the book he shows that all of physics can be condensed into a master equation

$$U = 0$$

(here left as an exercise for the serious student). Suddenly some limits of condensing physics with equations occur to the reader.

And this brings me to the point: *The Feynman Lectures on Physics* contain practically all the physics needed to do industrial research very well. My old physics teacher used to lament the fact that today's graduate students are experts in applying the Green's function method but get stumped if you ask them what the pressure is at the bottom of a swimming pool. There are lots of swimming pools in the application of physics. If you want to join me for a swim, you can float on *The Feynman Lectures*.

A lecture to a lady on Thomson's reflecting galvanometer

JAMES CLERK MAXWELL

The lamp-light falls on blackened walls,
And streams through narrow perforations,
The long beam trails o'er pasteboard scales,
With slow-decaying oscillations.
Flow, current, flow, set the quick light-spot flying,
Flow current, answer light-spot, flashing, quivering, dying.

O look! how queer! how thin and clear,
And thinner, clearer, sharper growing
The gliding fire! with central wire,
The fine degrees distinctly showing.
Swing, magnet, swing, advancing and receding,
Swing magnet! Answer dearest, What's your final reading?

O love! you fail to read the scale
Correct to tenths of a division.
To mirror heaven those eyes were given,
And not for methods of precision.
Break contact, break, set the free light-spot flying;
Break contact, rest thee, magnet, swinging, creeping, dying.

73

European atom

From *Time* (27 December 1937) p 28.

Lieutenant Colonel John Theodore Cuthbert Moore-Brabazon is a distinguished British authority on aeronautics, 'first Englishman to fly in England,' past President of the Royal Aeronautical Society. In his *Who's Who* entry, Colonel Moore-Brabazon lists his recreations as 'golf, tobogganing, yachting.' Last week he was engaged in another kind of recreation which took the form of a very pleasant altercation—not only typically British, but typical of the well-ballasted wit of the man of science anywhere—with Professor Edward Neville da Costa Andrade, FRS, FInstP, DSc, Quain Professor of Physics at the University of London, editor for physics of *Encyclopaedia Britannica*, author of *The Structure of the Atom, The Atom, The Mechanism of Nature*. Professor Andrade lists his recreations as 'golf, poetry, collecting old scientific books and useless knowledge.'

Colonel Moore-Brabazon is fascinated by the news of the atom's interior and behaviour which trickles out of the cloister into the writings of such interpreters as Jeans, Eddington and Professor Andrade. But he is also somewhat annoyed by the paradoxes and abstractions which result from the fact that atomic behaviour cannot be visualized or represented by commonplace physical analogy. In a letter printed by *Nature* last month he drew up a polite bill of complaint against physicists. A chief item was that after laymen have learned to regard protons, electrons, and other charged particles as nothing but electricity, the physicists adduce the neutron which has no charge and therefore cannot exist—although a stream of neutrons will knock the living daylights out of a block of paraffin. With equal politeness Professor Andrade replied, declaring in effect that it was really not the physicists' fault if atoms behaved in a way not explainable 'in anthropomorphic terms of likes and dislikes,' that physicists were not trying to be confusing but to obtain the best possible description of what Lord Rutherford called 'a world of its own'—the atomic nucleus. 'Now, perhaps' concluded the professor 'Colonel Moore-Brabazon will give me a logical statement of British foreign policy in the last ten years, which has puzzled me as much as the nuclear mechanics of the last ten years has puzzled him.'

74

Last week witty Colonel Moore-Brabazon came back with a reasonably logical statement of British foreign policy and a picture of Europe couched in the physicists' own terms, as follows:

'Europe may be looked upon as a nucleus composed of individual protons, not, however, all of the same size or power, mixed with a few neutrons of no charge and little mass. This is kept together by a strong force which prevents them flying apart, known as geography. This nucleus is not symmetrical as, included on its western edge, is a particularly powerful proton (Britain) that has 'wave characteristics' of a definite type peculiar to itself. In the south there is what might be called a neutrino (Italy). This also has, some think, wave mechanic aspirations. It is peculiar in this respect that its core is eternal (Rome) but its surround, some think, is ephemeral.

'Now the real trouble is that just as in the atom there are electrons in their orbits far away from the nucleus, so in this case there are colonies also revolving. These used to be attached, so to speak, to separate protons, but some years ago the nucleus was subjected to a terrific bombardment which shifted these electrons from belonging to one proton to another. One very powerful proton, in mathematical language generally designated thus 卐 suffered severely in this respect, with the result that the nucleus as such is no longer stable. It has been found, however, that if the western proton adds to its charge (by re-arming), although a state of strain between the two protons is introduced, the nucleus qua nucleus becomes more stable.

'I hope I have put this very difficult problem in simple terms for the physicists . . ..'

Royal Society of Chemistry. 1962 Christmas competition quatrain.

I'm grateful, Lord, for NMR,
*   For IR and UV,*
But may a technique ne'er be found
*   That will dispense with me.*

75

Error bars

JERRY D WILSON

One day I questioned Harold Knox, who was working on his dissertation, as to just what he was doing in the Neutron Lab. (They were using an old Cockcroft–Walton.) They always seemed to be a clandestine bunch over there anyway. He informed me that they made error bars. Of course, I asked to see a sample of their work, and shortly thereafter I received this letter:

Ohio University Neutron Lab
Ohio University
Athens, Ohio 45701
May 10, 1971

Dear Dr Wilson,

As per your request of May 10, 1971, I am enclosing a sample of the data points produced at the Ohio University Neutron Lab. The model (DP-1-EB) enclosed here is our standard point with error bar. This model is quite suitable for publishing. Data points with small error bars are also available; the cost per point, however, increases quite rapidly as the error bar size decreases. Data points without error bars (Model DP-1), our so-called ball park values, are available at reduced prices. Prices of our products are available on request.

Data point making and error bar reduction are old and cherished crafts. Much of the skill in these crafts has been passed down from generation to generation. Today with modern technology these crafts have reached a high state of perfection. We of the Ohio University Neutron Lab strive to maintain in our product the proper mix of Old World quality and modern technology in order that our products are the best available.

Yours truly,

Harold Knox

The British high energy physics community

Foreword (condensed) written by John Ziman, University of Bristol, for Jerry Gaston's book *Originality and Competition in Science* (Chicago: University of Chicago Press) 1973.

This book about my neighbours, the *uk-hep* and *uk-het* clans, contains much interesting information that will be used by other anthropologists in their analysis of our peculiar culture. The first point to make clear is that all members of these clans have dedicated their lives, by a series of magical ceremonies of great psychological weight, to a single purpose—the cultivation of the *elpar* fruit (*nucleonicus fermi*).

Despite the unworthy spirit of competition that occasionally seizes upon individuals and groups when some new variety is coming to seed or a new crop is ready for picking, the normal attitude is one of friendliness and cooperation between all members of the clans. All heps and hets of whatever tribe firmly believe that the cultivation of the elpar (which is, incidentally, quite inedible), is the most important thing in life, and that all the tribal resources should be devoted to this single end. Clannishness even extends across tribal boundaries. The hep and het clans are so distinct in their social roles that one may find closer fraternity between, say, the uk-hets, us-hets, eur-hets, and russ-hets, even though these belong to different tribes that are often at war with one another, than between the uk-hets and uk-heps, alongside whom they actually live.

The differentiation of role between heps and hets arises quite naturally out of the peculiar life cycle of the elpar. The reproductive phase, in which the seeds germinate rapidly, grow to young plants, flower, and produce a few new seeds, is under the charge of the hets; the fruiting phase, which lasts much longer, and leads to a large crop, is under the care of the heps.

The sole aim of the het is to breed a significantly new variety of elpar, as measured by the beauty of its flower and eventually by the fragrance of its fruit. This is essentially a solitary task, demanding no more than a few seed trays and pots, a little fertile compost, and a modest supply of water, light, and air. The secret of success is thought to lie mainly in the choice of ritual chants, which must be sung with perfect accuracy and clarity as the seed germinates and as the young plant pushes out its leaves and flower buds. The hets are, incidentally, the supreme masters of the symbolic language,

77

of seed raising. This explains both the division of labour between the clans and, to some extent, the subtle feelings of superiority and inferiority between them.

The uk-het and uk-hep clans date back to the beginning of this century, to the great headman JoJo and his adopted son Rŭth, who was himself the headman or adopted father of many of the older headmen of our day. Rŭth himself was what we should now call a hep, but he had an unusual knack for raising many sturdy new varieties of plant without the aid of ritual songs, and was indeed somewhat scornful of many of the more fanciful varieties being bred by the hets of his day. This attitude also, transmitted via his adopted sons, is by no means outmoded amongst the present generation of uk-hets, who still maintain amongst themselves something of the family feelings of the founder of their clan.

I must apologize to readers for this nonscientific description of the psychological and historical aspects of our tribal life, but it seemed that many of the observations recorded in these anthropological researches would lose their significance unless seen in context. As a humble member of a neighbouring clan who has had the good fortune to become acquainted with several visiting anthropologists, I felt that it might be of some use to say some of these things in my own simple fashion.

We really try to have only one new particle per paper.

PATRICK BLACKETT

If I could remember the names of all these particles I'd be a botanist.

ENRICO FERMI

80

Prepared by Susan Winarchick Cohick, Laura Kate Hutton, Janet Briggs and the Society of Physics Students, Pennsylvania State University.

All Power to Particles

By unanimous consent the Dummy Index Summation Convention hereby presents the following demands to the Department of Physics:

I: All charges against particles should be dropped immediately

II: Newton's First and Second Laws should be repealed, for the reason that they restrict the personal freedoms of particles

III: Newton's Third Law should be repealed since it is reactionary

IV: Maxwell's demon should be arrested and held without bail, on charges of restricting the normal (and tangent) ingress and egress of particles from their private habitats

V: The universal speed limit should be lowered, since travelling at such speeds invariably results in destructive collisions

The above demands are non-negotiable. If they are not implemented immediately drastic measures will have to be taken

The future European harmonic pendulum

Report of a working party appointed by The Arts Research Council

From *Orbit*, Journal of the Rutherford High Energy Laboratory, Didcot, England (December 1964) p 8.

Art today is on the threshold of far reaching discoveries which will come to fruition with a new theory of the nature of art and the fundamental processes of artistic expression. Before the advent of the Harmonic Pendulum, art was a haphazard, subjective and often highly emotional act on the part of the artist, but now we possess a powerful tool for the investigation of the very nature of Art itself.

Harmonic patterns have recently revealed an astonishing and unexpected richness in the form of over 100 recognisably different patterns. We now know that the majority of these are variations, 'excited states' as it were, of a few stable elementary patterns and to understand their properties we must soon go to still longer pendulums. Our present knowledge tells us that the increase in length, if it is to be significant, must be substantial. Progress is such that we shall need the new facilities in the early 1970s.

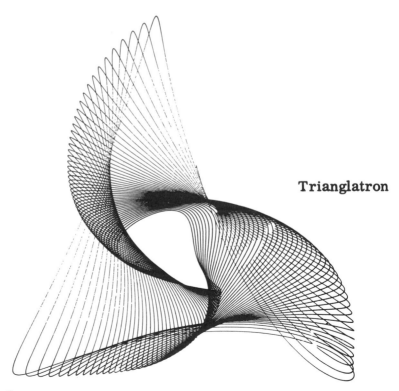

Trianglatron

Each substantial increase in pendulum length in the past has been amply justified. Election to the Royal Academy has always been won by going to a longer pendulum.

There are two proposed advances. One is to have a substantial increase in length and precision while at the same time achieving greatly improved writing speeds, i.e. an increase in the *intensity* of the patterns is of vital importance. To minimize the frictional effects of the pen, the mass of the pendulum (i.e. its kinetic energy) must be very large. We are therefore proposing a precision, high-intensity, high-energy

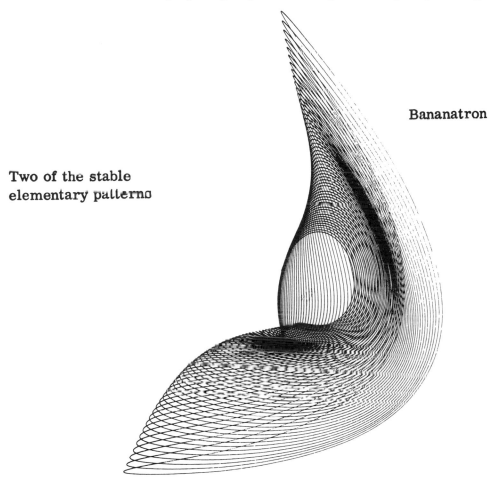

Bananatron

**Two of the stable
elementary patterns**

pendulum (HEP), 600 ft long, oscillating in a vacuum vessel 100 ft in diameter. All damping will be controlled by on-line computers.

The second requirement is to increase the pendulum length to the maximum possible extent without necessarily achieving a high intensity. To meet this urgent need we propose a 1400 ft pendulum suspended from the Eiffel Tower. Since the Tower itself is only 985 ft high, this would require a hole 415 ft deep to be dug at its base, but the resulting facility would enable European artists to compete on almost equal terms with the Americans. The American Congress has recently approved a project for a 1472 ft machine suspended from the Empire State Building. This device would provide information in the very long pendulum region where the most exciting developments are predicted.

The two proposals—the high intensity machine and the ultra-long pendulum project—are not alternatives but two complementary aspects of the same programme of high-energy pendulums, each in itself desirable.

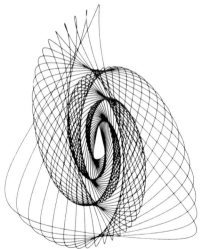

Heavy Depression

FINANCE AND MANPOWER
Experience indicates that only those countries with a massive home-based programme of HEP studies will derive maximum

84

benefit from the large international projects like those discussed above. It takes many years before a young artist is fit to make use of these large complicated machines and merely to ensure their efficient use, we have to insist that the first consideration be the wholehearted support of university art departments and the national colleges of art.

The total cost of these activities has been doubling every six months, so that by the early 1970s, we shall require £100 million per annum compared with the 19/6d per annum of the first pendulums.

We estimate that by 1970 there will be about 20 artists who will have to be accommodated elsewhere unless these new facilities are provided. Thus there is no difficulty in providing the necessary artistic manpower. However, a team of about 5000 engineers and technicians will be required to construct and operate the machines and we are concerned about the shortage of outstanding engineers for pendulum design. The qualities required, a grasp of structural engineering, wide experience in vacuum technology, and a strong head for heights, are rare in contemporary engineers. This appears to be connected with a deficiency in the basic training.

**Two of the unstable
excited states or
berserkatrons**

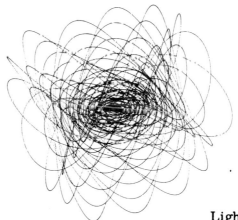

Light Hysterion

HIGH-INTENSITY MACHINE IN BRITAIN

It is very desirable that the 600 ft machine be built on British soil. There are several disused mine shafts of adequate depth which could be enlarged to accommodate the vacuum vessel and by reopening the railway lines we could have good accessibility, so valuable in large international projects. The presence of a large artistic community in our midst would be of enormous cultural value and the consumer spending of the staff, mostly foreign, would amount to a third of the UK contribution. A simple financial adjustment—for example a tripling of all salaries—could make the consumer spending equal the cost of the project.

SUMMARY AND CONCLUSIONS

The present generation of pendulums has provided us with a tantalizing picture of art. At the next stage of machine development, or the one after, we are likely to see all the art forms of past times unified by our knowledge of harmonic pictures. Quite apart from the exciting intellectual challenge of taking part in this great synthesis of artistic thought, the new picture of the nature of art is bound to have far-reaching effects on our whole way of life.

It seems inconceivable that a country which has been in the forefront of the World's artistic development for so long should not continue to take part in this most fundamental branch of the subject. Other European nations are determined to go forward. It would be a severe blow to the international development of art if we were not to remain with them and this country would certainly face a further increase in the art drain.

If the present programme is approved and the new 600 ft machine built in this country, it is likely that the requirements for HEP will be satisfied for a long time to come.

The artistic case for Europe continuing forcefully in this field is overwhelming; the equipment needed is technically feasible; the artistic manpower needed will be available; the money is trivial. Only conservatism or timidity will stop it.

The fluttering, spiralling flight of cyclotron evolution

DAVID L JUDD

Condensed from
IEEE Trans. Nuclear Science **NS-14** (4)
xiii–xx (1966).

My assigned task is to present something originally character-ized as an introductory and keynote address, containing a certain amount of historical background. I propose to review briefly what I regard as some of the important landmarks in the evolution of cyclotrons. You may now settle down for a lecture equivalent to well over 12 000 words, since I have prepared almost a dozen pictures, each being worth a thousand words. These pictures are designed to enable you to see the cyclotron through the eyes of others, so as to have a well-rounded outlook rather than a specialized one, and to appreciate the point of view of the various experts in this rapidly growing field.

The cyclotron as seen by . . .

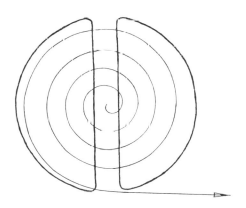

. . . the inventor

$p = 37.945067 \pm .00023$ MEV
0.03 × 0.05 cm;
± 0.000075 m rad.

. . . the experimental physicist

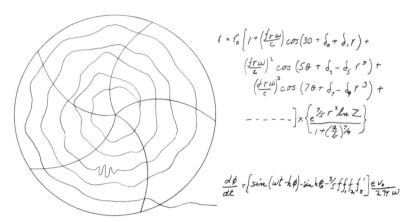

$$I = I_0 \left[1 + \left(\frac{frw}{c} \right) \cos(3\theta + \delta_0 + \delta, r) + \left(\frac{frw}{c} \right)^2 \cos(5\theta + \delta_3 - \delta_5 \, r^3) + \left(\frac{frw}{c} \right)^3 \cos(7\theta + \delta_7 - \delta_9 \, r^3) + \cdots \cdots \right] \times \left\{ \frac{e^{7/5} r^2 \ln Z}{1 + \left(\frac{a}{r} \right)^{7/9}} \right\}$$

$$\frac{d\phi}{dt} = \left[\sin(\omega t - h\phi) - \sin h\beta - \frac{3}{5} f f_1 f_2^{-1} f_3' \right] \frac{e v_0}{2\pi w}$$

. . . the theoretical physicist

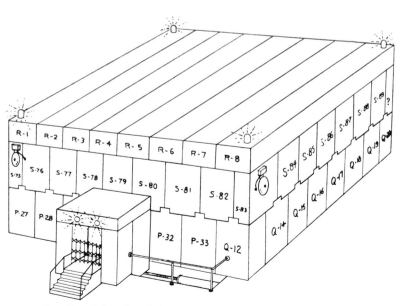

. . . the health physicist

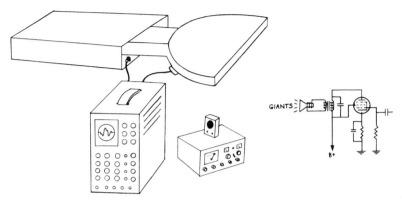

. . . the electrical engineer

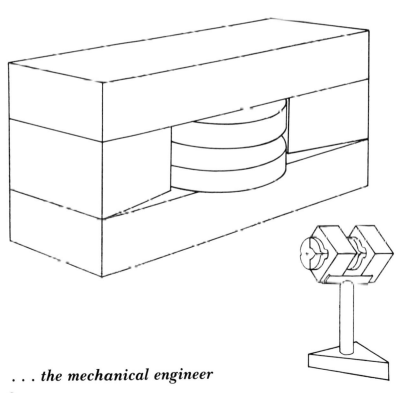

. . . the mechanical engineer

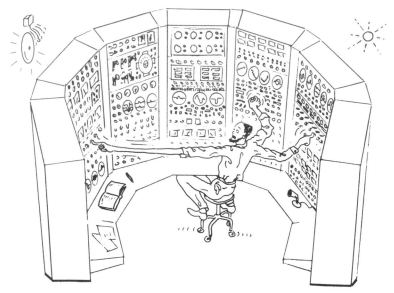

. . . the operator

. . . the laboratory director

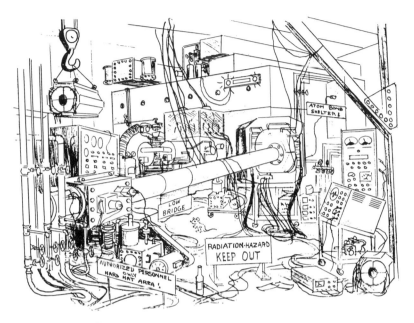

. . . the visitor

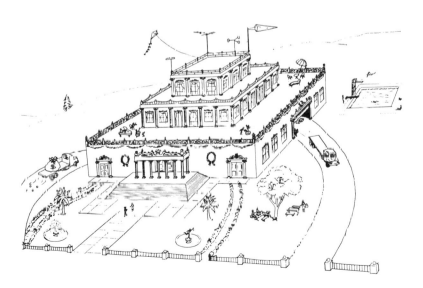

. . . the governmental funding agency

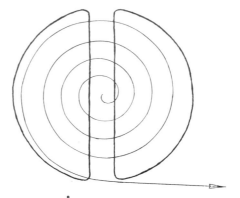

. . . the student

E O Lawrence, driver

From *Lawrence and Oppenheimer* by Nuel Pharr Davis (New York: Simon and Schuster) 1968, p 135.

Lawrence set so fast a pace, the Smyth Report notes dryly, that shrewd guesses had to take the place of adequate research . . ..

'We were supposed to put in sixteen hours a day, seven days a week and that was all. Lawrence considered the people in the Radiation Laboratory as extensions of his own hands and he didn't feel the need to keep them manicured' Carroll Mills recalls.

For the young, the roughest leadership can be effective so long as it remains nakedly honest. 'Once Lawrence ordered Dennis Gardner and me—Dennis later discovered artificial mesons and died of beryllium poisoning—to build a calutron power source' recalls Roger Hildebrand. 'He said we were to get it done in a day and a half. I consider it now a task that should have taken two people several weeks. But we didn't sleep and we improvised and in thirty-six hours we got the power supply going. Lawrence had guessed right about the time. That's where his genius lay—in estimating exactly what was humanly possible. I never knew him to lower a demand. . . . I remember when I was stacking lead bricks for shielding and Lawrence came and laboured beside me for a couple of hours. He didn't ask us to do anything he wouldn't.'

Ten commandments of electrical safety

From *Orbit*,
Journal of the
Rutherford High
Energy
Laboratory,
Didcot, England
(31 January 1965)
p 12.

(1) Beware of the lightning that lurks in an undischarged capacitor lest it cause thee to be bounced upon thy backside in a most ungentlemanly manner.

(2) Cause thou the switch that supplies large quantities of juice to be opened and thusly tagged, so thy days may be long on this earthly vale of tears.

(3) Prove to thyself that all circuits that radiateth and upon which thou worketh are grounded lest they lift thee to high-frequency potential and cause thee to radiate also.

(4) Take care thou useth the proper method when thou taketh the measure of high-voltage circuits so that thou doth not incinerate both thee and the meter, for verily though thou hast no account number and can be easily replaced, the meter doth have one and as a consequence bringeth much woe upon the supply department.

(5) Tarry thee not amongst those who engage in intentional shocks for they are surely non-believers and are not long for this world.

(6) Take care thou tampereth not with interlocks and safety devices, for this incureth the wrath of thy seniors and bringeth the fury of the safety officer down upon thy head and shoulders.

(7) Work thee not on energized equipment, for if thou doeth, thy mates will surely be buying beers without thee and thy space at the bar will be filled by another.

(8) Verily, verily I say unto thee, never service high-voltage equipment alone, for electric cooking is a slothful process, and thou might sizzle in thy own fat for hours on end before thy Maker sees fit to end thy misery and drag thee into his fold.

(9) Trifle thee not with radioactive tubes and substances lest thou commence to glow in the dark like a lightning bug.

(10) Commit thee to memory the works of the prophets, which are written in the instruction books, which giveth the straight info and which consoleth thee, and thou cannot make mistakes.

$h\nu$

G STEAD

From *Post-prandial Proceedings of the Cavendish Society*
[To the tune 'Men of Harlech'.]

All black body radiations,
All the spectrum variations,
All atomic oscillations
Vary as $h\nu$.

CHORUS
Here's the right relation,
Governs radiation,
Here's the new
And only true,
Electrodynamical equation;
Never mind your d/dt^2,
Ve or half mv^2
(If you watch the factor c^2)
's equal to $h\nu$.

Ultraviolet vibrations,
X and gamma ray pulsations,
Ordinary light sensations
All obey $h\nu$.

Even in matters calorific,
Such things as the heat specific
Yield to treatment scientific
If you use $h\nu$.

In all questions energetic,
Whether static or kinetic,
Or electric, or magnetic,
You must use $h\nu$.

There would be a mighty clearance,
We should all be Planck's adherents,
Were it not that interference
Still defies $h\nu$.

It does not take an idea so long to become 'classical' in physics as it does in the arts.

KARL K DARROW

Tea-time with Rutherford

From the 5th
Rutherford Lecture
of the Physical
Society, given by
A S Russell
(8 December
1950). Published
by the Physical
Society (London).

Rutherford loved chaffing his younger research workers on occasion at tea-time in his laboratory. As I was the only Scot amongst them I came in for a good deal of chaff on the real or supposed foibles of Scotsmen. He thought them an over-scholarshipped and over-praised lot. 'You young fellows come down here from across the border with such testimonials written by your Scots professors that, why man alive!, if Faraday or Clerk Maxwell were competing against you they wouldn't even get on to the short list.' One day he told us he had picked up a delightful phrase from one of the novels of H G Wells: 'a fellow of the Royal Society in the sight of God', and thereafter, on occasion, the young Scotsman of science was not only, on paper, miles better than Maxwell and Faraday, 'but already at twenty-four a fellow of the Royal Society in the sight of God'.

I once saw Rutherford genuinely surprised. We had been out on a laboratory visit to a works where one of the products was a plated glass teapot. (These were quite popular at one time.) The inner piece of glass held the warm tea, the outer faced the world. Between was deposited some white metal which shone brightly. It was as though a glass mirror had been fashioned into a teapot. To make polite conversation Rutherford asked the man showing him the teapot what metal was used for the plating. He was quite staggered when the man replied 'Platinum.' Platinum in those days was very rare and never spoken lightly of. Moreover, it happened that just at that time Rutherford had shown that what afterwards was called the nucleus of platinum had a charge of about one hundred positive units. Rutherford was consequently emotionally interested in platinum, and so expressed his incredulity about the answer given in a loud and sustained 'Wh-a-a-at?' He wasn't going to be fobbed off with sales talk or have his leg even lightly pulled. 'Platinum?' repeated Rutherford, 'Platinum? Platinum?' 'Yes' quietly answered the man, almost apologizing for the word. 'This isn't a case of "only the best being good enough" for these teapots. It's a case of the most expensive being the cheapest', and he started talking of thinness of films and coverage power and details of that sort. 'But do you really mean platinum?' asked Rutherford again, giving the man one more chance to withdraw. 'Oh, yes' said the man, quite cheerfully. 'It's platinum all right', and then, to bring it right down to the level of the Nobel Prizeman in Chemistry, he added: 'one of the osmium, iridium family of metals' and then, turning to me to clinch the matter, he said 'atomic weight 195.2.' The moment he said that, I knew, somehow, that it was platinum. Platinum, in fact, it really was. Rutherford, genuinely surprised but not yet defeated, then retaliated by asking how much was the cost of plating a single teapot. All he evidently wanted to know was whether it was five shillings, one shilling or what. It was clear from the way the question was received by the man that we were in for a long *viva*. 'Grade A or grade B?' he asked. Rutherford, at a venture, murmured rather boredly 'B'. 'Grade A is better' he was encouraged to believe. 'All right then' roared Rutherford 'grade A'. 'We don't actually quote for single teapots' resumed the conscientious man. 'I'll quote you the fiftieth of the cost of fifty.' 'All right' shouted Rutherford, whose patience had almost given out. 'Wholesale or retail?' continued the exasperating man, and there would have been a scene had not the answer to Rutherford's 'wholesale' put everyone in high good humour. For the answer that came at long last, given calmly and weightedly, as though empires depended upon it, was, believe it or not, twopence three-farthings! I can still hear Rutherford's long and loud laugh as he heard the unexpected news, and for some time afterwards he had a new story to work off on his friends.

The three Peter Zeeman memorial stained glass windows designed ▶ by Harm K Onnes in the Kamerlingh-Onnes Laboratory at the University of Leiden commemorating the discovery which became known as the Zeeman effect. The upper panel shows Zeeman making the original discovery as reported on 31 October 1896; the central panel pictures Lorentz and symbolizes the theoretical part of the work; and the lower panel shows the confirmation of the predictions on 28 November 1896. (Courtesy of Dr M Knook.)

The Space Window [overleaf], by Rodney Winfield, commemorates man's first landing on the Moon. At the centre of the window is a white glass circle where a piece of rock returned from the Sea of Tranquility, by the Apollo 11 astronauts, is set. (Courtesy of Washington Cathedral, Mount Saint Alban, Washington, DC.)

Pauli in Heaven

From Isaac
Asimov *Treasury of
Humour* (London:
Woburn Press)
1971.

Wolfgang Pauli was well-known for his quick penetration into the very essence of the theoretical analyses presented by his colleagues, and for his sure nose for error. It was consequently not at all surprising (the story goes) that God awaited his arrival, after his death in 1958, with anticipation.

'I presume' said God 'that there is much in the world of physics that puzzled you during your lifetime that you would be glad to have the opportunity of understanding now.'

'Yes, Lord' said Pauli 'for to tell the truth I am weary of watching my colleagues go wrong. For instance, I have always been disturbed over the fact that the proton has exactly 1836.11 times the mass of the electron though the electric charges are the same. Why so odd a multiple? Yet there must be a reason. It is just that all the theories I have seen which were designed to explain the matter were so ludicrously wrong.'

'Ah' said God. 'Here, then, in the language of twentieth-century quantum mechanics, is the explanation of the proton/electron mass ratio.' And he handed a sheaf of papers to Pauli.

Pauli looked through the sheets eagerly and rapidly, turned back to the first page, took a quick look at the fourth, and handed them all back to God.

'Still wrong!' he said with a sigh.

From *American
Journal of Physics*
45 422 (1977).

It was absolutely marvellous working for [Wolfgang] Pauli. You could ask him anything. There was no worry that he would think a particular question was stupid, since he thought *all* questions were stupid.

VICTOR WEISSKOPF

◄ The Albert Einstein Window and the John Glenn Window [overleaf], designed and made by Gabriel Loire of Chartres, France. Photographs by George Knight. (Courtesy of The Very Reverend David M Gillespie, Dean, Grace Cathedral, San Francisco.)

PHYSIKALISCHES LIED BY MOLLY KULE

A study on the diatomic scale, arranged for a cyanogen band.

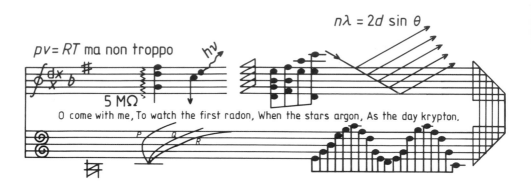

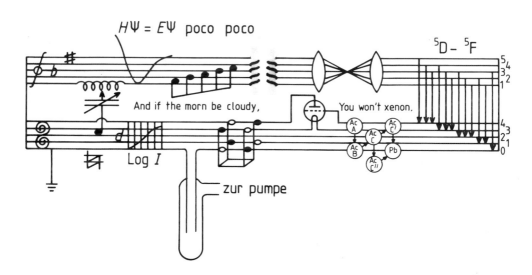

[This Lied is unusual in that it combines so many puns, wordplays, and physics items in one opus. It circulated at Princeton in 1942, possibly a work of the Princeton chapter of the American Institute for Useless Research. The illustration has been redrawn by Mrs Brenda Elliott of The Institute of Physics from copy supplied by Francis E Throw.]

Tracking the elusive quark

LEWIS GROSSBERGER

From *The New Yorker* (10 May 1976) p 32.

No man has ever seen a quark. How, then, do we know they exist? The circumstantial evidence is overwhelming. We have seen the incredibly tiny bubble tracks. We have heard the eerie cries in the night. (Not even a charging neutrino makes a sound like that.) And there are the ancient legends: the tales of human sacrifices to assuage the terrible anger of the quark gods; the chants of the shamans, who believed that quarks made the sun rise and fall, the seasons change, the rivers flow home to the sea; and that they were available in three colours and four flavours. We do not believe that, of course, and yet we know there is something out there . . ..

Dr Gaddis Quigley, of the Iota Institute, set out in June to track the elusive quark to its lair. I was pleased to be included in his party.

Dr Quigley was admirably frank in explaining the dangers inherent in our quest. Yet not a man among us—nor Eleanor Grommet, our physician and the only woman aboard, recruited at the last minute in response to picketing by a protest group—flinched at the prospect of facing the dread quark. All of us were qualified. Dr Quigley had hunted the snipe, the snark, and the carbuncle without qualm. I had once faced an excited hadron alone and unarmed, and spent my college vacations spelunking through black holes. We were all handpicked for calm under fire, and we would certainly need it while hunting the quark, with its unpredictable nature, its low boredom threshold, and its most uncanny natural defence mechanism—a life span of one-thousandth of a billionth of a second. 'This will make it extremely difficult to get off an accurate shot' Dr Quigley warned us. 'You will not have a second chance.'

Dr Filbert Cranshaw, of Scintilla University, our theoretician and public-relations man, was excited—perhaps overly so—at the possibility of our unearthing a 'charmed' quark. Recent sightings had been reported, but I and the others questioned their authenticity. Like such pop apparitions as the Indomitable Slow Man, the Big Mouth of the Rockies, and the Landlocked Nester, the existence of the charmed quark seemed more wishful than thinking.

After a week of provisioning and taking on hydrogen, we boarded our bubble chamber and moved out into the vast Subatomic Field, setting up a base camp in the forbidding Vector Meson. Our initial excitement was soon replaced by a pervasive unease, and, sheltered from the raging elements by our Quonsets, we spent a nervous first night. Jankowski played an endless quadrille on his ocarina until Chavann yelled for quiet. Dr Grommet went in to calm the boy, and was able to lull him to sleep by softly reciting the periodic table.

At dawn, we made directly for quark country, fanning out in a semicircle. It was a wild, luminous landscape, filled with lowing herds of muons, which scattered at our approach. Hyperons chattered overhead as we cut our way through the thick nuclei, and in the distance we could hear the mournful howl of the ever-dangerous lepton.

The hours passed, and we searched on until we were bleary-eyed. With dusk approaching, we finally turned back towards camp. Above, the quasars were pulsating dimly. Clumsily, I almost stepped on a sleeping proton and it sent up a shower of sparks. I was silently resolving to be more careful when suddenly I heard it—that unforgettable aching cry. It stopped me cold—the querulous, full-throated croak of the adult bull quark. Even though I'd only heard Dr Cranshaw's faculty-tea imitations, I recognized it instantly. Then came a metallic flash. Quigley was shouting, 'There! Behind the electron cloud.' I saw something move and squeezed off a laser beam. But I was rattled and it caromed harmlessly off a huge graviton.

Cursing my ineptitude, and thinking that probably we'd lost our quarry, I began running toward the spot where the flash had come from. Then, so quickly that I didn't realize what was happening, Soy Chavann was down, clutching at his toe and howling. The incredible was taking place. He was being squelched by a charmed antiquark, apparently the mate of the quark we'd just flushed. She was trying to protect her young.

I couldn't get off another beam, because Balbanian was between Chavann and myself and his Quantum Theorizer had jammed! It was young Jankowski, unarmed but for his ocarina, who got there first, and courageously flailed at the ground with his instrument. Too late. She was gone as soon as she had come, if not before, and Chavann lay shaking, a wan smirk on his lips.

'Cranshaw was right' he said in a horrible whisper. 'I had her in my microscope cross hairs, just for a millisecond. She exists—the charmed antiquark!'

'But now you must rest' ordered Dr Grommet, kneeling to give him a quinine shot.

As Dr Quigley came huffing up, Chavann beckoned for him to draw near. 'Leave me here' he said. 'I'm done for. You must return immediately to make the fall issue of *Particle Quarterly.*'

Dr Quigley looked stricken, but his voice was firm. 'Don't be foolish, son' he said. 'You're going to pull through. It's just a question of mind over matter.'

If anybody says he can think about quantum problems without getting giddy, that only shows he has not understood the first thing about them.

NIELS BOHR

Spin–orbit coupling

MARIA GOEPPERT MAYER

Contributed by Joseph E Mayer.

An explanation of spin–orbit coupling given to a 14-year-old girl [Marianne Mayer] on New Year's morning at about 5:00 am, at the end of a long, long party: 'Last night you danced a waltz. Now, the traffic of the ballroom goes around and around the room, and that is the orbit. In addition, each couple rotates about its own axis, and that is the spin. Everybody who has ever danced a fast Viennese waltz knows that it is much easier to spin the same way around as the orbit than the opposite way around. That is spin–orbit coupling.'

Reference frames

JOHN N HOWARD

From *Applied Optics*
19 (6) 827 (1980).
Philipp Frank was a good friend of Einstein's and visited him one time in the 1920s. Einstein was somewhat miffed at some of the public criticism of his relativity theory in the press and at widespread assertions that relativity was difficult to understand. He told Frank that he had just written a short easy book on relativity that any layman could understand: in fact his twelve-year-old daughter had read the book and understood it. A little while later, when Einstein had left the room, Frank turned to Ilsa and asked her if she had really understood her father's book. 'Yes' she replied 'all except what he meant by inertial reference frames.'

We recall another anecdote about Philipp Frank. One year he greeted his new class in philosophy of science by remarking 'The first topic we shall discuss concerns Space and Time. In what room should we meet and at what time?'

Space

FELIX BLOCH

From *Physics Today*
29 (12) 27 (1976).
There is another remark which Werner Heisenberg once made that I consider even more characteristic. We were on a walk and somehow began to talk about space. I had just read Weyl's book *Space, Time and Matter*, and under its influence was proud to declare that space was simply the field of linear operations.

'Nonsense' said Heisenberg 'space is blue and birds fly through it.'

This may sound naïve, but I knew him well enough by that time to fully understand the rebuke. What he meant was that it was dangerous for a physicist to describe Nature in terms of idealized abstractions too far removed from the evidence of actual observation. In fact, it was just by avoiding this danger in the previous description of atomic phenomena that he was able to arrive at his great creation of quantum mechanics.

The Einstein and the Eddington

[*Dr C A Murray has kindly contributed the following stanzas from the private records of the Royal Astronomical Society, prefaced by a letter by the author, W H Williams. The poem is here condensed.*]

Perhaps a short account of all that happened will interest you and serve as an excuse for its perpetration.

Eddington spent a couple of months with the Physics Department [Pasadena] in 1924. I was glad to share my office with him but to my disappointment found that he suffered from British uncommunicativeness in a most severe form. However, it happened one day that I asked him if he played golf and he acknowledged that he did. So I took him to the Claremont Club and thereafter we went there twice a week and played perhaps the worst golf ever seen on that course. Anyhow this broke the ice and there followed many pleasant talks and discussions, the memory of which has not faded.

On the eve of his departure, the Faculty Club gave him a dinner and Dean Walter Hart asked me to make a speech. After some efforts towards solemnity I descended to doggerel. Eddington, as you know, was an Alice in Wonderland fan. This and the allied topsyturvydoms of Carroll and Einstein, together with the irreverent way we both treated the royal game of golf furnished the motive of my 'poem'. I am glad to remember that it did amuse the guest of honour.

The sun was setting on the links,
  The moon looked down serene.
The caddies all had gone to bed,
  But still there could be seen
Two players lingering by the trap
  That guards the thirteenth green.

The Einstein and the Eddington
  Were counting up their score;
The Einstein card showed ninety-eight
  And Eddington's was more.
And both lay bunkered in the trap
  And both stood there and swore.

The time has come, said Eddington
  To talk of many things;
Of cubes and clocks and meter-sticks
  And why a pendulum swings.
And how far space is out of plumb.
  And whether time has wings.

I learned at school the apple's fall
  To gravity was due.
But now you tell me that the cause
  Is merely G$\mu\nu$.
I cannot bring myself to think
  That this is really true.

You say that gravitation's force
  Is clearly not a pull.
That space is mostly emptiness,
  While time is nearly full;
And though I hate to doubt your word,
  It sounds a bit like bull.

You hold that time is badly warped,
  That even light is bent;
I think I get the idea there,
  If this is what you meant:
The mail the postman brings today,
  Tomorrow will be sent.

If I should go to Timbuctoo
  With twice the speed of light,
And leave this afternoon at four,
  I'd get back home last night.
You've got it now, the Einstein said,
  That is precisely right.

And if, before the past is through,
  The future intervenes;
Then what's the use of anything;
  Of cabbages or queens;
Pray tell me what's the bally use
  Of Presidents and Deans.

The shortest line, Einstein replied,
  Is not the one that's straight;
It curves around upon itself,
  Much like a figure eight.
And if you go too rapidly
  You will arrive too late.

But thank you very, very much,
  For troubling to explain;
I hope you will forgive my tears,
  My head begins to pain;
I feel the symptoms coming on
  Of softening of the brain.

Einstein's philosophy of science

PHILIPP FRANK

From *Review of Modern Physics* **21** 349 (1949).

About ten years ago I spoke with Einstein about the astonishing fact that so many ministers of various denominations are strongly interested in the theory of relativity. Einstein said that according to his estimation there are more clergymen interested in relativity than physicists. A little puzzled I asked him how he would explain this strange fact. He answered, a little smiling, 'Because clergymen are interested in the general laws of nature and physicists, very often, are not.' Another day we spoke about a certain physicist who had very little success in his research work. Mostly he attacked problems which offered tremendous difficulties. By most of his colleagues he was not rated very highly. Einstein, however, said about him 'I admire this type of man. I have little patience with scientists who take a board of wood, look for its thinnest part and drill a great number of holes where the drilling is easy.'

Relativity

ROBERT WILLIAMS WOOD

'Observe' said Professor Einstein,
'These recent deductions of mine.
I've certain new notions
Of time, space, and motions
Which I shall proceed to define.'

'If we measure the length of a train
When in motion, my doctrines ordain
We shall find it much shorter
Than we really had oughter
Though I haven't the time to explain.'

'That space should go on without end
Is certainly nonsense, my friend.
I re-arrange EUCLID
And thus make him LUCIDE,
At least that's what I contend.'

'Suppose an elliptical pill
Is blown at high speed through a quill.
It should really appear
In the form of a sphere—
And I'm perfectly certain it will.'

'To Sir Isaac, an apple that fell
Offered problems on which I won't dwell.
But a rising dumb-waiter
Has made my fame greater
And explains gravitation as well.'

It seems probable that the grand underlying principles of physical science have been firmly established and that further advances are to be sought chiefly in the rigorous application of these principles to all phenomena.

ALBERT MICHELSON

R W Wood—improviser

[*Professor Wood was noted not only for the skill with which he performed significant experiments in optics and spectroscopy but also for his propensity to construct apparatus from odds and ends. The following account reflects this approach.*]

THE 40-FT SPECTROGRAPH

From *Philosophical Magazine* Ser. 6 **24** 677 (1912).

The discovery of the satellite lines which accompany the resonance lines, and the change in their position and intensity which resulted from changes in the distribution of the intensity within the exciting line, made a careful photographic study of the absorption spectrum of the iodine and of the emission spectrum of the mercury arc much to be desired. I accordingly fitted up during the summer a plane-grating spectrograph of 12.5 metres focal length. As this spectrograph appears to be, with the possible exception of Professor Michelson's ten-inch grating instrument, the largest and most powerful in the world, a description of the method of mounting may be of interest. The grating is a plane one ruled by Dr Anderson on the 15000 machine, and the photographs which I have taken with it of the iodine absorption spectrum show that it yields its full theoretical resolving power of 300000 in the fourth order, which is bright enough to yield a fully exposed photograph of the solar spectrum with a very fine slit in three minutes, when used with a lens of 12.5 metres focus. My laboratory at East Hampton is in an old barn of very large size, and as temperature differences and striae due to air-currents can be better avoided by having the entire apparatus under cover, I first mounted the apparatus in the building. I soon found, however, that vibrations of the building due to wind, and probably expansion and contraction of its frame due to the changing position of the Sun, made it very difficult to secure satisfactory photographs, though I occasionally obtained one showing full theoretical resolving power. I finally determined to mount the instrument entirely independent of the building, supporting the grating and the lens on a cast-iron pier outside the building, and the slit and plate-holder on a similar pier sunk in the ground just inside the wall of my dark-room. Not wishing to order a lens until I had tried out the instrument under the conditions prevailing at East Hampton (I had some fear of vibration resulting from the surf on the beach a quarter of a mile distant), I borrowed, through the courtesy of Professor Campbell and the regents of the University of California, a very fine six-inch achromatic lens of 40-ft focus. The pier, which carried this lens and the grating, I improvised from two six-inch water mains which were slightly damaged by last winter's frost, and were to be had for the asking. The joints of these pipes had been ground together, and the two were bolted together as when laid for service. The resulting pipe was sunk to a depth of six feet in the ground, and a triangular brass bracket from my junk heap, which had originally formed a truss on an old fashioned support for a large reflecting telescope, was bolted to a bent piece of wrought iron, which was in turn bolted to the flange on the top of the water-pipe. The cell of the lens was fastened to the brass triangle with adjusting screws in the usual manner. The grating was mounted on a circle taken from a discarded spectrometer, which was turned by a worm gear. A bevel gear removed from a discarded hand-drill, was fastened to the worm, the small gear-wheel being turned by a long rod, made by fastening four lengths of cheap brass-covered iron curtain rod together. This rod passed through the wall of the barn, terminating in a wheel a little to one side of the plate-holder and slit. When photographing with the apparatus, the small gear-wheel is thrown out of mesh with the large one by sliding a wooden support through which the rod passes a little to one side, the object being to prevent the transmission of vibrations from the grating-house and the long wooden tube to the grating support. The wooden tube, down which the light from the slit passes to the lens and grating and back again to the plate or eyepiece, was supported on posts and roofed over to prevent it from being heated by the Sun. The end of the tube passes through the wall of the grating-house, but does not come in contact with the lens support. The plate-holder and slit were supported on a shorter piece of water-pipe, which passed through the floor of the laboratory without contact with it. The camera consisted of two wooden boxes, one sliding within the other (for focusing), joined to the end of the long tube with black cloth, which shut out the light but did not transmit vibrations. The long tube was made by nailing eight-inch boards together, and was painted black on the inside. Some trouble was given by spiders, which built their webs at intervals along the tube, a difficulty which I surmounted by sending our pussy-cat through it, and subsequently destroying the spiders with poisonous fumes.

Schuster's library

ROBERT L WEBER

The University of Manchester's eminent physicist Sir Arthur Schuster (1851–1934) had wide-ranging interests in physics and applied mathematics; he carried on a world-wide correspondence with other scientists. He amassed a remarkable collection of reprints of scientific papers. Many were autographed and many contained marginal queries and comments which would be valuable to a historian of this period of exciting development in physics.

Shortly after the end of World War II, a catalogue offered for sale the Schuster collection of reports, complete with oak bookcases. In this time of shortages of many things, including metal shelving, the Librarian of a large US state university found it expedient to buy the Schuster library. He found use for its bookcases; the collection of scientific papers was dispersed and destroyed.

In these days, a man who says a thing cannot be done is quite apt to be interrupted by some idiot doing it.

ELBERT HUBBARD

Two-, Three-, and Four-Atom Exchange Effects in bcc ^{3}He

J. H. Hetherington and F. D. C. Willard

Physics Department, Michigan State University, East Lansing, Michigan 48824

(Received 22 September 1975)

We have made mean-field calculations with a Hamiltonian obtained from two-, three-, and four-atom exchange in bcc solid ^{3}He. We are able to fit the high-temperature experiments as well as the phase diagram of Kummer *et al.* at low temperatures. We find two kinds of antiferromagnetic phases as suggested by Kummer's experiments.

Everything about the paper abstracted above is serious except that the second author is a cat (note his signature). Professor Hetherington explains: 'I had prepared the paper, now called Hetherington and Willard, and was rather proud of the work, considering it suitable for rapid publication in *Phys. Rev. Lett.* Before I submitted it I asked a colleague to read it over and he said "It's a fine paper but they will send it right back." He explained that this is because of the Editor's rule that the word "we" should not be used in a paper with only a single author. Changing the paper to the impersonal seemed too difficult now that it was all written and typed; therefore, after an evening's thought I simply asked the secretary to change the title page to include the name of the family cat, a Siamese called Chester, sired one summer by Willard (one of the few unfixed male Siamese cats in Aspen, Colorado). I added the initials F D in front of the name to stand for *Felix Domesticus* and thus created F D C Willard.

'Why was I willing to do such an irreverent thing? Against it was the fact that most of us are paid partly by how many papers we publish, and there is some dilution of the effect of the paper on one's reputation when it is shared by another author. On the other hand, I did not ignore completely the publicity value, either. If it eventually proved to be correct, people would remember the paper more if the anomalous authorship were known. In any case I went ahead and did it and have generally not been sorry. Most people are amused by the concept, only editors, for some reason, seem to find little humour in the story.

'When reprints arrived, I inked F D C Willard's paw and he and I signed about 10 reprints which I sent to a few friends.

Two of these reprints had some later consequence. One official at NSF keeps one in his office and when the conversation lulls with one of his visitors he takes it out and tells the story. Since most of his visitors are seeking funds, I presume they all think it very funny if he does.

'I had always secretly hoped that Willard would get some kind of invitation to speak on his work. I later learned that he probably would have received such an invitation had it not been for one of the "signed" reprints. The reprint was sent to a young French physicist. He was in a meeting choosing invitees for the LT-15 conference, when someone suggested that they "invite Willard, he never seems to get invited anywhere." The young physicist said he was not sure, but he thought Willard might be a cat. He brought the reprint to the next meeting and passed round the copy, which said "Compliments of the authors" followed by our two signatures. It may or may not be significant that I did not receive an invitation to that conference either.

'The paper in *Recherche* [No. 114, September 1980, p 972] signed by F D C Willard occurred after some disagreement among the authors about the details presented in that popularization. Willard, being already published in the field, seemed a reasonable pseudonym for the authorship—no one could blame a cat for getting a few details wrong! We can also note that his time spent learning such excellent French explains his rather sparse publication record.

'The story has now been told many times and my wife can add that she sleeps with both authors!'

It is also a good rule not to put too much confidence in experimental results until they have been confirmed by theory.

SIR ARTHUR EDDINGTON

Captain DeKhotinsky

PAUL E KLOPSTEG

From 'Potpourri
and gallimaufry'
Science **140** 594
(1963).

At the turn of the century, there was a coterie of young men in Chicago who occasionally 'went out on the town' for a good time. Most prominent among these was A A Michelson, who seemed to have an affinity for individuals with great skill in optical and mechanical craftsmanship. The other four were such craftsmen: Edward Petididier, an optical technician; Albert Porter, constructor of fine apparatus; William Gaertner, founder of the Gaertner Scientific Corporation, who had been the mechanical 'right hand' of Samuel P Langley and who had built the first powered aircraft for Langley which should have flown but failed to do so for other than aerodynamic reasons; and Achilles DeKhotinsky, a former captain in the Russian navy, who had come to America and become instrument maker at Ryerson Laboratory during the gay nineties.

When I joined Central Scientific Company in 1921 to take over responsibility for its research and development along ·with production, I found that I had 'inherited' as a development engineer none other than Captain DeKhotinsky, then 71 years old. He was a man of aristocratic bearing and pride almost to the point of arrogance, completely disdainful of workmanship which failed to measure up to his exacting standards. He was himself highly skilled in all the known laboratory arts, and had ruled some of the first six-inch diffraction gratings. 'The Captain,' as he was known around the organization, had many other accomplishments about which he liked to talk when I had time to listen. He showed me a Belgian patent issued to him on a pasted storage battery plate which predated the Planté process. Among scientists his name was probably best known for the laboratory cement which he had invented, and to which his name was attached. His speech and diction bore a strong flavour of his native tongue. Among his irreverent associates in the organization, the cement was known as 'Tsementski.' Its composition and production were still secret some 40 years after its invention, shared only with a stepson who produced it and with CENCO the exclusive distributor.

According to The Captain, the cement was invented because of necessity. He needed a sealing material for making electric light bulbs which he had used in street lighting in Vienna several years before the invention of the carbon-filament lamp in Edison's laboratory. DeKhotinsky's bulbs used a thin graphite rod as the incandescent element, hermetically sealed in a vacuum with the cement. Later the cement proved itself so useful in various applications that it became a standard supply item in many laboratories.

As a captain in the Russian navy in the late 1870s, DeKhotinsky had been assigned to supervizing and expediting the construction of a couple of battleships for which the Tsar's government had contracted with an American firm in the New York area. This tour of duty gave him his first contact with the United States, and, being observant and inquisitive, he used his spare time to visit places and see things that interested him. One visit was to Menlo Park, NJ, to Edison's laboratory. Edison was absent, but an assistant showed the visitor around, and described what was being done in trying to develop means for producing light by incandescence produced by an electric current. Platinum wire was being used. DeKhotinsky, having in mind his bulbs with graphite rod elements, said 'Vy don't you use car-r-rbon?' He told me that several weeks later a patent application was filed on the carbon-filament lamp, in which the filament consisted of carbonized organic material such as thin bamboo strips. He liked to believe that his remark about carbon gave Edison the idea of a carbon filament. Perhaps it did.

A sequel to this story is that the Edison patent or patents were sold to the General Electric Company which improved the product and developed the production process for making the bulbs. DeKhotinsky had by this time completed his navy assignment, had received commendation and a jewel-encrusted watch from the Tsar for his good services, and decided that his future lay in America. He immigrated in the early 1880s and started an incandescent lamp factory at Marblehead, Mass. After he had begun production, economic lightning struck in the form of a patent infringement suit, and his manufacturing venture had to be closed out. The experience, as well it might, seemed to have left him a permanent scar of embitterment.

When I first knew DeKhotinsky, in 1921, he was very much 'the cat that walked by his lone.' He shunned those in the organization whose status he regarded as inferior to his, which included most of the personnel; but he seemed to enjoy visiting with those whose authority he recognized. He made designs and sketches of experimental models, and the shops constructed them, but never to his satisfaction. His stock comment was that they 'bugger-r-red' his ideas. He was obviously not too happy in his work.

About 1924 the physics department at the University of Michigan, under the leadership of my old friend H M Randall, began a research program in fine structure in the far red and infrared. This created a need for precision-ruled diffraction gratings to produce maximum intensities in the regions selected for study. Randall enquired whether I could put him in touch with an instrument maker who could build a ruling engine and produce the gratings needed.

Although The Captain was already 74 years old, it seemed a made-to-order position for him; it would give him release from surroundings which were obviously not congenial and challenge his skill in a way which could warm his enthusiasm and stir his pride in accomplishment. I recommended him to Randall. He was hired, went to Ann Arbor, and there spent a number of years in great contentment, doing precisely what he liked to do, and giving satisfaction to the group for which he worked.

The Doppler effect for sound waves was tested by the Dutch meteorologist Buys-Ballot in an endearing experiment in 1845—as a moving source of sound he used an orchestra of trumpeters standing in an open car of a railroad train, whizzing through the Dutch countryside near Utrecht.

STEVEN WEINBERG in *The First Three Minutes*

My anecdotage:

Apocryphal tales

RONALD D EDGE

It seems as though eminent physicists invite apocryphal tales to be told about them. Ever since Archimedes jumped out of his bath shouting '*Eureka*', people have assumed that physicists are, to say the least, odd. For example, Leonardo da Vinci shared the magnificent megalomania of the Renaissance, devising projects beyond the technical capabilities even of today. Newton's peculiarities were of a different kind. He was an introspective recluse who solved problems by intuition and later justified them. A story is told of how he informed Halley of one of his most fundamental discoveries of planetary motion. 'Yes' said Halley 'but how do you know that? Have you proved it?' Newton was taken aback. 'Why, I've known it for years. If you'll give me a few days, I'll certainly find you a proof of this', as in due course he did.

Absent-mindedness seems to be a justifiable attribute of many physicists, and numerous amusing situations result. For example, J J Thomson was notoriously vague. He also patronized garage sales. One day his wife woke up in their bedroom at Trinity College to find that Thomson had left for work but that his trousers were still on the chair by the door. She dashed to the phone and called the porter at the lodge to stop Thomson leaving college without his trousers. In this case, however, the alarm proved unnecessary—Thomson had picked up a secondhand pair of trousers at a garage sale the previous day.

Absent-mindedness also plays its part in a tale told about Paul Dirac, who is married to Eugene Wigner's sister. The story goes that Dirac was noticed wandering around Cambridge in a somewhat distracted fashion, and when questioned he replied 'Wigner's sister is having a baby.'

There are a number of humorous incidents associated with Ernest Rutherford. It is said that while conducting an oral examination he asked the student his favourite question, 'What is the self-inductance of a wedding ring?' The student immediately replied '654.3, Sir'. Rutherford answered 'Oh, and what units are you using?' Said the student 'Arbitrary units, of course, Sir.' The early 1930s was a period of great growth at the Cavendish Laboratory. At that time, the reaction of high-energy deuterons impinging on a deuteron target was not yet understood. Mark Oliphant was working on this problem, and he told me that for weeks they were unable to fathom out what was going on. Then, about two o'clock one morning, Rutherford burst in saying 'I've got it! I've got it!' and finally the enigma was unravelled. J Robert Oppenheimer was working in Rutherford's lab at that time and it is often forgotten that he went there as an experimentalist. However, one day Rutherford came into the lab to find Oppenheimer beating on the floor next to his equipment with a hammer. It seems that Oppenheimer had been so frustrated by his experiment that he had to release his tension and he didn't want to take it out on the equipment which was too precious. After this episode, Rutherford decided that Oppenheimer would do better as a theorist.

There must be a semi-infinite number of tales related about Richard Feynman. The first seems to have originated at Los Alamos where he was irritated by papers being locked away in safes overnight, and he would come in, pick the lock, take the papers out and leave a little note in the safe saying 'Feynman was here.' In his room he had a board on one of the walls covered with partially smoked cigars. When the wife of one of the members of his group had a baby, he felt that the least that he could do was to smoke the cigar he was offered, and what was left before he became sick he fastened onto the board. It is a well known fact that his wife sued for divorce on the grounds that he would go down and play bongo drums at night while thinking about a physics problem, a practice that drove her mad.

It is told of Sir Harold Jeffreys, the famous theoretician and geophysicist, that he was at one time consultant to an oil company. Attending their meeting in London, Sir Harold sat quietly while the company's scientists discussed their problems and difficulties. About half way through the morning, they stopped, thought and said 'What do you think of this, Sir Harold?' 'I think it's time for coffee' he said. The rest of the morning was spent with more of the scientists' discussion. Then, 'What do you think, Sir Harold?' 'I believe it's lunch time' was the reply. The same happened in the afternoon— except that it was tea time. Came the end of the meeting, and the scientists finally said 'Well, now that you've heard it all, Sir Harold, what is your opinion?' 'I'm glad it's your problem and not mine' was his concluding remark.

Some physicists extract a great deal of amusement from practical jokes. R W Wood was one such. He was interested in photography and wished to take a fish-eye view of a certain bridge, but every time he set up his equipment passers-by would stop to look at it. So finally he painted the whole apparatus a brilliant red, set it up and then ran from the scene as if it were about to explode. Everyone, of course, followed him away from the experiment, which then proceeded to take the picture minus observers. Another tale is told of both Wood and Norbert Wiener, who are reputed to have put a very heavy gyroscope into a suitcase. The gyroscope was capable of being spun by a rope which came through a hole in the top of the suitcase, and just before getting off a train, Wood is supposed to have started the gyroscope spinning, and then handed the suitcase to a porter on alighting. All went well until Wood suddenly turned a sharp corner. The porter tried to follow suit, and the case rose vertically up in the air, causing great consternation to all present.

Wiener was famous for his talks on cybernetics, and on one occasion while discussing the problems with some students one of the students came up with a question and Wiener, in order to point out the statistical improbability of such a thing ever happening, said 'Why, that's as improbable as a bunch of monkeys having typed out the Encyclopaedia Britannica.' The student thought for a second and said 'But that's happened once, anyway.'

A marvellous tale is told about Kerr Grant, of the University of Adelaide. He had a large steel ball hung from the roof by an electromagnet, which at some point during the demonstration was switched off so that the steel ball fell with a resounding thud into a bucket of sand placed underneath. However, one day the students moved the bucket just sufficiently that the ball missed the bucket, leaving a jagged crack in the concrete floor. The following year, Grant carefully marked an X with chalk on the floor where the ball should land, and placed the bucket over it. However, this time the

117

students did not move the bucket, but instead they erased the chalk mark and drew another about a foot away. Grant came in, saw the chalk mark, carefully moved the bucket over it (smiling as he did so)—and turned off the electromagnet.

The diabolical heating scheme of J C Maxwell

or,

Maxwell sums nae sae bonny

JOHN LOWELL

Come in, come in, so ye're the mon whose supernatural skill
Will circumvent the Second Law and keep me frae the chill!
Ye cost me awfu' dear, ye ken—ma ain immortal soul—
But I'll no fret, I'll surely save an awfu' lot on coal.

  So here's your bat, ma cannie friend, and here's the little door
I've made for ye in ma hoose-wall, that ye maun stand afore
And watch the fleetin' molecules and, when I say, begin
Tae keep the slow and cauld anes oot, and let the hot anes in.

  Whit's that?—Ye canna tell the fast anes frae the slow?
And ye a michty daemon tae! Hoots, mon, don't ye know
The Doppler shift? Och, hark ye weel, I'll tell ye whit tae do—
Just bang the bluidie-reid anes back, and let the blue yins through!

  Och aye, that's richt, I had forgot, it's ower cauld, that's richt,
For molecules tae radiate appree-ciable licht;
And onyway, I ken it noo, it widna work at a',
Thanks tae that interfeerin' Kirchhoff's Thermal Law.
But dinna fash—this wee sma' torch, wi' ultraviolet ray
Will let ye see the molecules and keep the cauld away.

  That's grand—that's muckle braw—ye're doin' fine the noo.
But wait, there's something gane agley—whit are ye tryin' tae do?
Ah ken it fine ye need the torch tae enable ye tae see—
But dae ye need tae generate sae muckle entropy?
Ye're makin' more than ye destroy—O Jings, can it be true
The Second Law's inviolate e'en tae the likes o' you?

118

Proof by induction

EDGAR W KUTZSCHER

Professor Nernst liked to 'spice' his lectures with many interesting stories, anecdotes, and jokes. We, as students, appreciated this habit tremendously and I remember that even his first assistant, when he had to take over the lecture in the absence of Professor Nernst, used to say, 'At this particular point, Professor Nernst likes to tell such-and-such a story.' One story I remember went something like this:

Nernst, after discussing the first and second fundamental laws of thermodynamics, used to say 'As you remember, the *first* fundamental law of thermodynamics was formulated and theoretically proved by *three* physicists, Robert Mayer, James Prescott Joule, and Herman von Helmholtz; the *second* law of thermodynamics was formulated by *two* great physicists, Rudolph Clausius and William Thompson; and I developed the *third* law alone. So, you see, first law by three, second law by two, third law by one—which proves a fourth fundamental law of thermodynamics cannot exist.'

Stauffer's law of the entropy of change

Contributed by Don Stauffer.

It is obvious by observation of natural processes that pocket change possesses entropy, and follows the laws of thermodynamics. Thus: 'The entropy of coins is inversely proportional to their denomination. That is, pennies possess maximum entropy, silver dollars minimum. Therefore, change in a pocket will spontaneously decay, in a manner to increase the entropy of the system (pocket), towards a state of all pennies. The change will never spontaneously reform into a state of higher denominations (lesser entropy).'

This explains why, unless intervention is made, a pocket full of change, will, after several monetary transactions, decay into a pocket full of pennies.

First get your facts; then you can distort them at your leisure.

MARK TWAIN

International Court of Scientific Nomenclature
Rival claims of scientists:
Celsius vs McKie
Before the Lord President of the International Court

A S McKie in *NPL News* **230** 10 (21 June 1969).

The plaintiff, Anders Celsius, applied to the court for an order restraining the defendant, Alex McKie, from publicly disputing the right of the plaintiff to commemoration in the form of the naming of a unit after him.

Counsel for the plaintiff said that it had come to the plaintiff's notice that on 15 May 1969 the defendant had expressed the view that the unit of temperature should be called the Degree Centigrade rather than the Degree Celsius, thereby denying the right of the plaintiff to commemoration. His client was an eminent man, more so than the defendant, and had from 1730 to 1744 occupied the chair of Astronomy at the University of Uppsala. He had made extensive studies of the question of the measurement and expression of temperatures at a time when that question was wide open and had chosen with great foresight a metric solution to the problem. The rightness of his choice was vindicated in 1948 by an International Commission which decreed that thenceforward the unit of temperature should be called after him. The defendant's objection was meretricious, for it was undeniable that he wished to avoid the effort required of all scientists to give the plaintiff his due. Other units of temperature were called after equally eminent scientists and the use of emotive and metaphysical terms such as Absolute was to be deplored.

Appearing on his own behalf, the defendant said that the claim of the plaintiff should be opposed on two grounds: firstly that of historical precedent and secondly that of qualification.

On the question of historical precedent, the defendant said that in 1887 the Deutsche Anatomische Gesellschaft undertook to systematize the nomenclature used by anatomists which at the time verged on the chaotic. The Commission appointed in 1889 reported in 1895, saying that they found great divergence of opinion on the question of the inclusion of personal names and had decided to retain only those so firmly established in the language and affection of anatomists that they could not be left out. They reduced the number of anatomical terms from some 50 000 to 5528 and in the process reduced the personally named anatomical features to 146, commemorating 103 anatomists, in the *Basle Nomina Anatomica* of 1895. The *Jena Nomina Anatomica* of 1937 and the *Nomina Anatomica Parisiensa* of 1955 reaffirmed the principles that anatomical terms should be primarily memory signs but should preferably have some informative or descriptive value and that eponyms should not be used.

On the question of fitness for commemoration the defendant said that it was surprising that a Professor of Astronomy should be remembered only for his work on thermometry, and it was ironic that the plaintiff should be elevated to the scientific pantheon to take the place of his far more distinguished compatriot Anders Jonas Ångström. The irony was made clear when one remembered that although Celsius was Professor of Astronomy in the University of Uppsala it was Ångström, who had been called the father of modern spectroscopy, who made the greater contribution to astronomy by his studies of solar spectra when he was Professor of Physics at the very same university and whose book on the subject remained the standard work for nearly twenty years. The plaintiff could not even claim to be a pioneer of thermometry for that distinction should surely go to de Réamur, 'the Pliny of the Eighteenth Century,' who had been compared to Francis Bacon and who was so far ahead of his time as to believe that the scientific and the useful are indissolubly linked, a novel idea even today.

HIS LORDSHIP chose to disregard all arguments as to the fitness of the plaintiff for commemoration, and found that he could not improve on the findings of the Anatomical Commissions of 1895, 1937, and 1955. He noted that although the earliest of these, no doubt out of deference to the practice of the time, did not insist that eponyms be avoided at all cost, the position had changed by 1955 when the recommendations were put much more strongly. He remarked in passing that the commemoration of scientists resembled in kind but not in degree the process of canonization, in which the claim of a candidate was subjected to a most rigorous investigation lasting sometimes for several centuries. He took the opportunity to express satisfaction that the secular system of units was in its last decline and wondered that the literary world had allowed a person of such disputed eminence as Pound to be commemorated when there were other poets with undeniably greater claims. He was pleased that the deplorable precedent created for the animal kingdom by commemoration of the Ounce would soon be abolished. Who knew if, unless the principle of the avoidance of eponyms were not firmly established on this occasion, some eminent scientist might not claim commemoration not only for himself but also for his equally famous pedigreed dog, Bonzo? The order would not be granted.

---

mµ—an opticist's lament

M J MERRICK

From *Applied Optics* **8** (7) 1371 (1969).

The tears do flow
The dirge doth play
Millimicron
Hath passed away.

I met her first
In High School Science
Hardy and Perrin[1]
Sealed the alliance.

Alas, She's gone
Her soul hath flit
To join Her love
The Ångstrom unit.[2]

I see Her face
I hear Her cry
Four hundred times[3]
In the violet sky.

Her form doth haunt
Me, though unseen;
Five hundred times[4]
In the ocean's green.[5]

Her name I murmur
When lights are low
Six hundred times
In the embers' glow.[6]

My friends, they say,
Forget your Milly
Nanometer's
Better, rilly.

Well off with the old
And on with the new!
(Millimicron
I still love you.)

[1]In the literary famine of the late 1930s, one of the Beautiful Books.
[2]Late lamented.
[3]Poetic licence; the author's vision actually cuts off at 457.
[4]More or less.
[5]Flotsam's Bay, low tide.
[6]Milford's patented Briquettes at 1200 K.

The King who kept his 0 °C

EDWARD CROWLEY

From '*Dissent*' *The Observer Magazine:* (20 January 1974).

Once upon a time there was this King. He was on the run from his Loyal Opposition, who were anxious to penetrate his chain-mail with a yard (0.914 m) of cold steel. One day he had a close shave (0.03 mm) when a group of Loyal Opposition knights stopped and asked the way to Runneymede. 'First left, second right, and if you stop at the King's Arms have a pint (0.568 litres) on me' he replied. 'Obviously a peasant' concluded their leader, the Black Knight.

The King was glad to see the back of them. 'It's about time for my dream' he mused. At 10 o'clock (22.00 hours) on the dot he always had this dream about a magic weapon against which armour was useless—I'd like to see your armour stand up to a 6 in (152.4 mm) anti-tank gun at 10 paces. But this time Lady Guinevere, with her 38–20–38 (96–50–96) figure, took over his dream. When she started to divest herself of her wimple and things, the scandalized monarch half-heartedly protested, 'Desist. This is a respectable dream.' But he added prudently, being nobody's wiseacre (0.405 wisehectares), 'See me in my cave tomorrow at three.'

Alas! She never turned up and he was reduced to watching the antics of his pet spider vainly trying with every ounce (0.21 hectogram) of its strength to spin a web over a horrifying 2 ft (0.61 m) sheer drop.

'Don't you ever get tired of doing that?' he demanded irritably. The spider replied as quick as a flash (299 300 km s^{-1}) 'At least I didn't let the Black Knight half-inch (1.27 cm) my kingdom.' Stung, the King narrowed his royal orbs. 'Bet you the salvage rights to my Crown Jewels in the Wash against the manufacturing rights of Gossamer Spun Yarn you'll never bridge that gap.

'You're on' said the spider.

The King knew there wasn't a grain (0.65 decigram) of truth in that hoary rumour about the lost Crown Jewels. He knew where those Crown Jewels were, and they certainly weren't in the Wash. He felt miles (kilometres) better and his gloom disappeared as if a ton (1.016 tonne) weight had been lifted from his mind.

Even when the crafty insect claimed its pound (0.454 kilogram) of flesh after sub-contracting the job to a firm of web erectors who spanned the gap with Easiweb lightweight alloy, the King kept his cool (0 °C). He let the spider have the useless salvage rights and promptly stopped dreaming about anti-tank guns. Impressed with the spider's business acumen, he determined henceforth not to yield an inch (2.54 cm) to fate and without more ado embarked on the liberation of his kingdom from the oppressors.

He rustled himself a one horse-power (0.7457 kW) white charger with only 20 000 miles (32 187 km) on the mileometer (kilometreometer) and galloped over to the castle of the de James brothers. With the help of these second-generation Norman outlaws, plus a modicum of rigged elections and personation and a soupçon of blackmail he soon made the citizens see things his way. During all this the Black Knight was fully occupied trawling the Wash for some Crown Jewels which, rumour had it, were lost there.

Finally, the King, never a man to hide his light under a bushel (0.36 hectolitre) sold the film rights of his story to Twelfth-Century Fox. With the proceeds, he got his Crown Jewels out of hock and then made it impossible for Lady Guinevere not to marry him (I refuse to divulge the method) so I don't see why he shouldn't live happily ever after.

Name for unit radian frequency

M STRASBERG

From *Journal of the Acoustical Society of America* **41** 1367 (1967).

Now that the term 'hertz' is gaining rapid acceptance as the name for unit cyclic frequency in cycles per second, it is time we considered a name for unit radian or 'angular' frequency. It seems natural to choose 'avis' as the name for unit radian frequency in radians per second. This pair of names will point up the fact that the choice of one or the other to indicate frequency is purely a matter of personal taste. Moreover, the German-sounding *hertz* would be nicely balanced by the French-sounding *avis*.

Those of us who prefer to express frequency in units of avis rather than hertz admit freely that we are second, but we believe that the use of avis will make things much easier.

On the mathematics of committees, boards and panels

BRUCE S OLD

From *Scientific Monthly* **63** 129 (1946).

[In this paper the author adopts a mathematical approach to analyse the efficiency of functioning of committees, boards and panels. Detailed equations are derived and graphs plotted to determine such relationships as actual versus calculated work output; efficiency versus numbers; efficiency versus intelligence; intelligence versus rank; etc. The extract below describes one of the mathematical functions which appears in the equations.]

There are some conditions which can ruin the efficiency of a committee even though it may have intelligent members and a competent chairman. The most serious of these is the heckler-saboteur function, $f(hs)$. The main types of heckler-saboteurs, many of which are well known, are as follows.

(a) The normal man. (All *Homo sapiens* have some faults.)

(b) The jolly fellow who is always 22 minutes late and then holds up proceedings 7 more minutes telling (off the record) his latest joke. Although he never does any committee work, he is such a good egg he never fails to get reappointed.

(c) The man with an elephantine memory to whom all ideas are old and who can still quote all the reasons used to turn down any one of them 11 or 12 years ago. No chance for viewing anything in a new light is given—the mere fact that he had heard of it before is sufficient reason in his mind to vote against any idea.

(d) The man who is against initiating any work, because, since there is a shortage of scientific manpower, any new project undertaken is bound to interfere with the progress of all existing projects (particularly two of his pet programmes).

(e) The poor fellow sent to represent his boss who has instructed him in such a manner that all he can do is sit there and say, 'I don't know', or 'I have no authority to speak for my agency'.

(f) The policy man who is afraid the rest of the committee is trying to take away some of his power and authority. Thus, he views each question not from the standpoint of whether it is the best thing to do, but whether the answer given might possibly be misinterpreted by anyone as permission for someone to infringe upon his cognisant empire.

(g) The man who is on the defensive. He suspects the committee has been formed just to change (Note: for the better) his method of doing something.

The 'legal' value of π

M H GREENBLATT

From *American Scientist* (December 1965) p 427A.

Many people have heard, at some time during their schooling, that 'some legislature, somewhere, once tried to legislate the value of π, and set it equal to 3.' The very idea of trying to legislate upon something as unlegislatable as the value of π is ludicrous. But such a bill was actually considered, and the details of this and the way it was handled, culminating in its rejection by the State Senate, can be somewhat amusing.

The bill in question is House Bill # 246, which was introduced in 1897 into the Indiana State Legislature. The bill was introduced by Representative T I Record, representative from Posey County. It does not suggest a single number for the value of π but rather suggests several different numbers. The bill was presumably offered as a contribution to education in the State of Indiana.

The first part of the bill states that the area of a circle is equal to the area of a square whose side is $\frac{1}{4}$ the circumference of the circle. If we represent the radius of the circle by r, the circumference would be $2\pi r$, and the bill would have us believe that the quantity $(2\pi r/4)^2$ is equal to the area of the circle. We have been taught that the area of the circle was simply πr^2. The area suggested in the bill would be correct if we assumed that π was equal to 4. The bill then goes on to mention that the ratio of the chord to the arc of 90° is as 7 to 8. We would state that the chord of 90° in a circle of radius r, was equal to $r\sqrt{2}$, and the arc of 90° is simply $(\pi/2)r$. This latter 'truth' would lead to a value of $\pi = \sqrt{2} \times 16/7$. (This last statement is closer to the truth than was the first statement.)

The same paragraph goes on to say that the ratio of the diagonal to one side of a square is as 10 is to 7. The assumption here is that the square root of 2 is exactly equal to $10/7$. This approximation is good to 1%. The bill then says that the ratio of the diameter to the circumference of a circle is as $5/4$ is to 4 (or, $\pi = 16/5 = 3.2$). The paragraph in question winds up by stating that, 'since the rule in present use fails to work, it should be discarded as wholly wanting and misleading in the practical applications.' The bill ends with the triumphant statement that the author has 'solutions of the trisection of the angle, duplication of the cube, and quadrature of the circle, which will be recognized as problems which have long since been given up by scientific bodies as unsolvable mysteries, and above man's ability to comprehend.'

When the bill was first introduced into the House of Representatives, in Indiana, it was referred to the Committee on Swamp Lands. The person who referred the bill to this committee is not known, but if he were known today, he might be honoured for having given such a diplomatic appraisal of the worth of the bill.

The Committee on Swamp Lands apparently recognized that the bill was not really in their province, and they recommended that the bill be considered by the Committee on Education. The Committee on Education considered the bill, reported it back to the House, and recommended that it should pass. And it did pass, unanimously, 67 to 0.

In the Senate, the bill fared a little bit worse. It was referred to the Committee on Temperance! (Perhaps the same shrewd chap who referred the bill to the House Committee on Swamp Lands had a part in referring it to the Committee on Temperance. A wonderful choice of committees!) The bill passed the first reading in the Senate, but that is as far as it ever went. After that first passage, the senators were properly coached, and on the second reading, the Senate threw out this 'epoch-making discovery' with much merriment.

We need education in the obvious more than investigation in the obscure.

OLIVER WENDELL HOLMES

Because mathematicians get along with common words, many amusing ambiguities arise. For instance, the word *function* probably expresses the most important idea in the whole history of mathematics. Yet, most people hearing it would think of a 'function' as meaning an evening social affair, while others, less socially minded, would think of their livers.

E KASNER and J NEWMAN

Science in double dactyls

By letter from Alan Holden, formerly at Bell Laboratories.

Higgledy, piggledy
William R Hamilton
Wrote on mechanics, in-
Frequently joked,

Wrote on quaternions,
Tackled intractable
Hydrodynamical
Problems, and croaked.

Contributed by the author, Jay M Pasachoff

Higgledy-piggledy
C P 0 5 3 2
Pulses its crabbiness
Here from afar.

Is it a message from
Extraterrestrial
Little Green Men on a
Dense neutron star?

[*Cambridge Pulsar at 5 hours and 32 minutes of right ascension (celestial longitude) is the name for the pulsar in the Crab Nebula. When pulsars were first discovered it was thought that they might be signals from Little Green Men, and the first four were known at first as* LGM 1 *through* LGM 4.]

By letter from A P French ex Cavendish Laboratory; now at Physics Department, Massachusetts Institute of Technology.

It occurred to me that you might like to see the few scientific double dactyls that I wrote during a wave of enthusiasm for this genre in 1966. My complete collection was annotated (somewhat tongue in cheek) for non-scientific readers.

Higgledy piggledy
Leopold Kronecker
Made a small delta[1] and
Said, with great glee,

'Unmathematical
God made the integers—
Then left the hard work to
People like me.'[2]

128

Higgledy piggledy
Werner K Heisenberg
Said 'I'm uncertain[3], but
Pretty astute.

'Give me a chance at a
Multidimensional
Matrix and Bingo! I
Anti-commute.'[4]

Higgledy, piggledy
Erwin 'H' Schrödinger[5]
Said 'Here's a recipe—
Don't ask me why

Structures of atoms are
Fully described by a
Quantum-mechanical
Function called ψ.'

Higgledy piggledy
Robert A Millikan
Scrutinized oil-drops and
So measured e;[6]

Studied phenomena
Photoelectrical,
Won the Nobel Prize and
Said 'Look at ME!'[7]

---

[1] The famous Kronecker delta: $\delta_{ij} = 1$ for $i = j$, $= 0$ for $i \neq j$,

[2] His actual remark was: 'God made the integers, all the rest is the work of man.'

[3] Remember his uncertainty principle.

[4] The reader will doubtless recall the quantum-mechanical anticommutation relations of the type $pq - qp = h/2\pi i$.

[5] Fraudulent. He had no middle name—but 'H' makes a good middle initial in view of its symbolic association with statistical mechanics (H) and quantum theory (h).

[6] 'e' is (of course) nature's atomic unit of electric charge.

[7] According to some reports he was not noted for his modesty.

129

Hymne to Hymen

[Dedicated by Blanche Descartes to Hector Pétard on the occasion of his marriage.]

From *Eureka*, Journal of the Archimedians, Cambridge University Mathematics Society no. 17 (October 1954), pp 5–7.

The wind was blowing soft, the sun was sending
along their space–time geodesics wending
millions of photons, orange, green, and yellow,
making the scene enchanting, warm, and mellow,
as by diffraction or reflection they're diverted
into the eye, and so to sight converted.
* Their energy, or frequency, is reckoned*
* in million kilocycles to the second.*

Love, from his pedestal in Piccadilly
is duly energized, and willy-nilly
a billion arrows sends with high velocity . . .
for he's a chap of great precocity . . .
* a billion? . . . more, in the vicinity*
* of undenumerable infinity.*

The photons, $h\nu = E$ obeying,
set many orbital electrons swaying.
The arrows, which ignore Dirac's equations,
cause sinusoidal cardiac palpitations
* in every youthful bachelor and spinster*
* within a neighbourhood of old Westminster.*

Hector Pétard, that well-known big-game hunter,
feared that his intellect was growing blunter.
A variation problem had him really nettled,
he couldn't see how it could well be settled:
who was it maximized charm, wit, and grace,
and was the fairest of the human race,
* yet minimized, under those same conditions,*
* all sorrowful and nasty dispositions?*

An arrow, flying straight without deflection,
with Hector Pétard's heart made intersection
(as well it might, with probability p,
there's arrows almost everywhere, you see).
Sensing a sudden break in his dejection,
he gave a glance in a north-west direction
(direction cosines θ, θ, o,
where $\theta^2 = \frac{1}{2}$). Our hero
saw there a lovely maiden, smiling gaily,
reading the Telegraph, or some such daily.
   Seeing his look, she too felt quite elated,
   and so his greetings were reciprocated.

Her conversation gave great satisfaction,
they had a strong Newtonian attraction.
She was kind-hearted, generous, and thrifty,
her IQ very much above 150,
her dazzling figure, when it was in focus,
beat hollow any algebraic locus,
her legs enveloped gracefully in nylon.
She solved his problem to within ε.
   And when his feelings he had deeply sounded,
   he found his love for her was quite unbounded.

'You are my true reciprocal' quoth he, . . .
'And you my contragradient' said she. . . .
'My converse, dual, polar, transposition,
my only isomorph in disposition,
my image, inverse in the sphere of life,
for future time why not become my wife?' . . .
'Your company is very sweet communion:
I think our meet ought to include our union.' . . .
'The union is contained within the meet
only of equal sets. The proof is neat.
   Come, charming one, let's be identified:
   I'll be the bridegroom, you the blushing bride.'

Our couple at this moment hail a carriage.
'Hey driver, speed to church. We want a marriage.'
The vicar seeing Hector questioned whether
they're certain he and she would live together
connectedly as long as they drew breath,
and never separated but by death?

'Oh, yes, we're positive. Oh, absolutely,
and swear by Harold Jeffreys resolutely;
this is no deviation due to chance,
it has statistical significance.'
So while he handed him the golden torus,
the vicar, quickly marrying them before us,
explained, as he performed the operation,
'This is an irreversible transformation.
You, Hector, owe to her in calm or storm,
convergence absolute and uniform,
    by involution you're uniquely mated,
    in fact, harmonically conjugated.'

Then Hector turned to her after the mating,
and ceased to oscillate, but osculating,
he said 'Beloved one, now you are mine,
and I am ever yours, this ring's a sign
of an implicit perfect right ideal.
Of us two conjugates, the sum is real,
the difference is pure imaginary,
and ever negligible it shall be.
    We two are definitely integrated,
    and never shall be differentiated.'

Like logs, on adding them they multiply.
No conservation law can here apply,
but step by step, proceeding by induction,
a joyful family is in construction.
    Best wishes from their friends in Trinity
    mount steadily towards infinity.

The moral of this episode is sweet.
Their hearts in resonance together beat.
In harmony, and equal in persuasion,
they form a Biharmonical Equation.
But though this romance has us thrilled, enraptured,
we're not quite sure the lion's really captured.
Sometimes we wonder if, when we confront her,
we'll find the lion's swallowed up the hunter.

Resolution of the paradox
A philosophical puppet play

ABNER SHIMONY

From *Zeno's Paradoxes* ed. Wesley C Salmon (Indianapolis: Bobbs-Merrill) 1970.

DRAMATIS PERSONAE: Zeno, Pupil, Lion
SCENE: The school of Zeno at Elea.

PUP. Master! There is a lion in the streets!

ZEN. Very good. You have learned your lesson in geography well. The fifteenth meridian, as measured from Greenwich, coincides with the high road from the Temple of Poseidon to the Agora—but you must not forget that it is an imaginary line.

PUP. Oh no, Master! I must humbly disagree. It is a *real* lion, a *menagerie* lion, and it is coming toward the school!

ZEN. My boy, in spite of your proficiency at geography, which is commendable in its way—albeit essentially the art of the surveyor and hence separated by the hair of the theodolite from the craft of a slave—you are deficient in philosophy. That which is real cannot be imaginary, and that which is imaginary cannot be real. Being is, and non-being is not, as my revered teacher Parmenides demonstrated first, last, and continually, and as I have attempted to convey to you.

PUP. Forgive me, Master. In my haste and excitement, themselves expressions of passion unworthy of you and of our school, I have spoken obscurely. Into the gulf between

133

the thought and the word, which, as you have taught us, is the trap set by non-being, I have again fallen. What I meant to say is that a lion has escaped from the zoo, and with deliberate speed it is rushing in the direction of the school and soon will be here!

The lion appears in the distance.

ZEN. O my boy, my boy! It pains me to contemplate the impenetrability of the human intellect and its incommensurability with the truth. Furthermore, I now recognize that a thirty-year novitiate is too brief—*sub specie aeternitatis*—and must be extended to forty years, before the apprenticeship proper can begin. A real lion, perhaps; but really running, impossible; and really arriving here, absurd!
PUP. Master . . .
ZEN. In order to run from the zoological garden to the Eleatic school, the lion would first have to traverse half the distance.

The lion traverses half the distance.

ZEN. But there is a first half of that half, and a first half of *that* half, and yet again a first half of *that* half to be traversed. And so the halves would of necessity regress to the first syllable of recorded time—nay, they would recede yet earlier than the first syllable. To have travelled but a minute part of the interval from the zoological garden to the school, the lion would have been obliged to embark upon his travels *infinitely* long ago.

The lion bursts into the schoolyard.

PUP. O Master, run, run! He is upon us!
ZEN. And thus, by *reductio ad absurdum*, we have proved that the lion could never have *begun* the course, the mere fantasy of which has so unworthily filled you with panic.

The pupil climbs an Ionic column, while the lion devours Zeno.

PUP. My mind is in a daze. Could there be a flaw in the Master's argument?

Saving time: a commentary

NORMAN SPERLING

From *Journal of Irreproducible Results* (1976) p 29.

Among the greater absurdities perpetrated on us is 'Daylight Saving' Time. Fearful of the night, politicians think they postpone it by renaming the hours. Calling '6 pm' by its alias '7 pm' changes only the schedules of people already independent of the Sun. Cattle still moo by solar time, and each day's sunlight ration is unaltered.

Back home in Michigan 'Daylight Saving' Time is itself postponed till April. But Michigan's long tradition opposing clock-tinkering is not a true answer. Michiganders merely sink out of synch with eastern merchants.

When the Sun is up, our intensely technological society shields it into submission. In its absence, we light artificially. We have but trivial use for 'saved' daylight.

Astronomers *need* darkness, however. Let us press for DARKNESS SAVING TIME. Rename '6 pm' as '5 pm' or even '4 pm'. Dusk would come sooner and make night-watching more convenient.

Clocked schedules would be unaffected. Farmers will welcome the dawn getting up to meet them. If clock-tinkering is to be done, Darkness Saving Time makes the most sense.

But any renaming of hours that retains antique Standard Time Zones treats symptoms, not causes. Invented a century ago for telegraph and railroad companies, zones serve areas of a then-useful size. Transoceanic commerce has transformed a nineteenth-century blessing into a twentieth-century headache. Jet lag knocks days out of intercontinental business men. Zones tangle East-West communications. Technology has out-moded the zone.

Any Standard Time that serves the whole Earth must be fair to all longitudes. Global Hawaii Standard Time would condemn European business to perpetual darkness. Expanding Central European Time from 3 zones to 24 would do the same to Oceania. Choosing any local solar time gives antipodes permanent night-time business days.

However, Solar is not the only kind of time. A day is how long Earth takes to turn once. We mark the spins by the Sun's reappearance at some standard position—hence 'High Noon.' Smoothing out minor inequalities (astronomers call their sum the 'Equation of Time') we get 'Mean Solar Time.' The standard is 'Greenwich Mean Time', at 0° longitude near London. A mean solar day has 24 hours, each with 60 minutes, and so on.

A mean solar day takes 3^m56^s longer than Earth needs for a spin, however. Consider that Earth orbits 360° around the Sun in about $365\frac{1}{4}$ days. You can see that each day we are about a degree farther around our orbit. Earth takes 3^m56^s of extra spinning to face the Sun each day.

If we don't worry about a Sun we usually ignore, we could skip that 3^m56^s. We would then use Earth's rate reckoned by the remote stars. This is called the 'Sidereal Day' and has 24 hours of sidereal time. Every year has 366 sidereal days.

The obvious standard time for our globe-circling technology is therefore MEAN GREENWICH SIDEREAL TIME.

Losses from zone differences, jet lag, and short years will disappear. A workday would be about a minute shorter, so there would be no change in work expectations. But with a 366th day 4 times as often, the extra day would increase production . . . unless the politicians declare a holiday.

Everyone, world-wide, will get fair daylight and darkness each year. No longer will astronomers always be out of phase. Many months yearly our prime working hours will be part of the normal workday.

People will always rise with the same stars. They may even learn a few constellations. Planet motions along the Zodiac will be readily apparent. Citizens thus introduced to their celestial heritage may get curious, and dramatically increase planetarium and astronomy club attendance.

Mechanical clocks can easily be adjusted to run 4^m per day faster (my watch often does already). Electric clocks would still use 60 Hz, though the hertz would fit the sidereal second.

At last everyone in the world would work at the same time. We'd all communicate better. Commerce would be more efficient and productive. Only agriculture, tied to plants, animals and the solar day, would be left out. Almanacs would give sidereal-to-solar conversions so cows could be milked at the proper time.

Modern transportation and communication make MGST practical. But the situation was foreseen two centuries ago in *The Astronomers' Drinking Song* (verse 11, to the tune of *Yankee Doodle*):

How light we reck of those who mock
  By this we'll make to appear, sir,
We'll dine by the Sidereal clock
  For one more bottle a year, sir!
But choose which pendulum you will,
  You'll never make your way, sir,
Unless you drink—and drink your fill,
  At least a bottle a day, sir!

Could the earth run backward?

From *Man and Time* by J B Priestley (London: Aldus Books) 1964 It has been pointed out that, if we can imagine observers moving away from the Earth at speeds comparable to that of light, some curious time effects would appear. If they moved at exactly the speed of light, let us say, on Christmas Day, any Earth light signals would remain the same so that they would, so to speak, stay in Christmas Day. But if they moved faster than the speed of light (a most unlikely event), the time would appear to be reversed for them, because they would overtake the signals sent out by Christmas Eve and then by the day before Christmas Eve, so that Earth time would appear to be running backwards. Speeds faster than that of light can be imagined, but are not likely to be achieved, except in thought. But if thought moves, what is its speed?

Which reminds one of the old limerick:

There was a young lady named Bright
Whose movements were faster than light;
  She went out one day
  In a relative way
And returned on the previous night.

Twixt earth and sky with matter-horns

RICHARD H LYON

From *Journal of the Acoustical Society* **32**, 1162 (1960).

Impedance matching is a technique which nature applies to the solution of many of her problems and man has learned from her to do as well. Her greatest triumph is probably the design of the middle ear but this note is concerned with one of her greatest failures, the matter-horn.

A typical matter-horn is shown in figure 1. The precise function of these devices has baffled many naturalists but it has long been clear to the author that they were intended as impedance matching devices to allow seismic energy to be released from the earth's crust and be dissipated harmlessly in the upper atmosphere. This harmful energy, which in its more concentrated forms is associated with San Francisco and Agadir, has even more insidious results at lower levels. Constant bombardment of the lower extremities by these rays can lead to fallen arches, varicose veins, and pelvic fatigue. In short, seismic radiation is killing us!

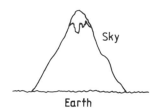

FIGURE 1. A typical matter-horn.

It was to rid the earth of this problem that nature produced matter-horns. In the final assembly, it appears that a grievous error was committed. As the author pointed out to some of his colleagues in 1945, matter exists in three stages: gaseous, liquid, and solid. The matter-horn must match the high impedance of the earth (solid) to the lower impedance of the sky (gas) and therefore, the mouth of the horn should point upward, not downward. Fortunately, it may not be too late to rectify the situation and save unborn generations from the awful effects of seismic radiation.

The modification of present matter-horns which the author proposes may be seen in figure 2. It is a simple and straight-forward application of the principles of pure science. Observing that a certain amount of cooperation between scientists

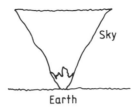

FIGURE 2. The proposed modification.

and engineers will be desirable (along with the usual congressional appropriation), may I suggest that a study group be formed to see how this project might most fruitfully proceed. I am sure the members of the Society will give these proposals the attention they most obviously deserve.

Courtesy of Dr Max Schatzmann, Lucerne.

Astronomical funding

Translated by Colena Jordan from *De Sterrekunde en de Mensheid* by M Minnaert (Katwijk: Servire BV) 1946.

The observatory at Poelkowo was from the start one of the best establishments in the world, because the Russian government had expended on it an amount unheard of at that time. When the Tsar came to see the completed institute he was conducted around by F G W Struve, the first director. At the end of the tour, the Tsar asked 'Well, Struve, are you satisfied?' This very careful astronomer, who did not want to deprive himself of further money, answered, 'For the time being, your Majesty.'

A short time later King Friedrich Wilhelm of Prussia visited Poelkowo and was amazed at the excellence of the building and the instruments. He exclaimed repeatedly, 'Wonderful, beautiful! If only my Bessel had such facilities; if Bessel could only see this!' On his way back to Germany he decided to visit 'his' Bessel at Königsberg. There he saw even more how inadequate the instruments were. He gave Bessel not additional money but a medal, because with such primitive instruments he had attained such excellent results.

Jeanne Hopkins in *Isaac Asimov's Science Fiction Magazine* (August 1980), p 6.

Higgledy piggledy
Albert A Michelson
Did his experiment,
*   Came away miffed:*

'Need a more accurate
Interferometer—
Back to the drawing board—
*   Can't get the drift.'*

O, telescope!

O, telescope, instrument of much knowledge, more precious than any sceptre! Is not he who holds thee in his hand made king and lord of the works of God?

JOHANNES KEPLER in *The Dioptrice* (1611)

Planetoid for sale

H H TURNER

From *Astronomical Discovery* (1904) pp 26–7, quoted by R A Oriti (1980) *Griffith Observer* **44** (8) 20.

Johann Palisa (1848–1925), an Austrian astronomer, was an exceedingly successful discoverer of minor planets, having found 121 of these objects during his lifetime. It is the discoverer's prerogative to name the planetoid, and so it must have been challenging for Palisa to find names for all of his discoveries.

Palisa once saw how the right to name minor planets might help him finance an eclipse expedition. In the astronomical journal *Observatory* for 1885, he published the following advertisement: 'Herr Palisa being desirous to raise funds for his intended expedition to observe the Total Solar Eclipse of August 1886, will sell the right of naming the minor planet No. 244 for 50 pounds.' The offer was not, however, snatched right up, and in the meantime Palisa found two more minor planets. Soon thereafter, however, his patience was rewarded. He proudly announced in the same journal: 'Minor planet No. 250 has been named "Bettina" by Baron Albert de Rothschild.' Baron Rothschild was delighted to name the object after his wife, and Palisa was happy to have the opportunity to observe the eclipse.

Celsius–the astronomer

Based in part on *A Biographical Dictionary of Scientists*, ed. T I Williams (Chichester. John Wiley) 1969, p 102.

Anders Celsius (1701–1744) of Uppsala, Sweden is not widely known for his astronomical accomplishments, which include brightness measurements of stars, systematic observations of the aurora borealis, and, in collaboration with others, a measurement of the length of a degree of latitude in Lapland. His chief claim to fame, however, resides in his convincing scientists to adopt a centigrade scale of temperature measurement. For his original version, he assigned 0 to the boiling point of water and 100 to the temperature of melting ice. A year later, Linné and Strömer reversed this designation and established the centigrade scale generally used for many years in scientific measurements. In 1969 the General Conference of Weights and Measures, choosing to honour scientists by attaching their names to units, decreed that the centigrade scale, as righted by Linné and Strömer, should be called the Celsius scale.

The satellite question

From *Punch*, reprinted in *The Observatory* **640** 281–4 (1927).

At an extraordinary meeting of the Solar Branch of Celestial Bodies Ltd, the proposed Satellites Disputes Bill came up for discussion.

The Sun having occupied the focus, the minutes of the last meeting were read by the secretary (Mercury) and duly approved.

The Sun rose and said he wished to call their attention to the unsatisfactory conduct of certain of the satellites. These, he complained, were in the habit of interposing periodically between their primaries and himself, with the result that he suffered eclipse and lost prestige and dignity. This eclipse policy, he continued, had been deliberately adopted by the Satellites' Union. There could be no excuse for it, as, with the whole of the heavens at their disposal, satellites might easily arrange their orbits so as to avoid this annoyance.

He wished to make his own position as luminous as possible. His business was to provide central heating and lighting for the whole solar system. He endeavoured to give satisfaction. Personally he resented any attempts, whether organized or not, to interfere with the execution of his duties. He would instance the total eclipse programme which had been attempted on 29 June. He had recently received many complaints from the Earth that on that planet they were not receiving their due quota of his attention. Much as he regretted his inability to be always with them—he had many claims upon his time—he thought his critics might see to it that, when he did shine, he should be allowed to do so unmolested. (Applause.)

The Earth, speaking in support of the proposals, said it was high time that legislation was introduced to restrict the motions of satellites. His own satellite, to whom reference had just been made, had recently been the subject of considerable apprehension in the Astral Plane. In brief, he was never certain what she might do next. It might not be common knowledge that the Moon was gradually parting company with him. Her general conduct was most erratic and irregular, and only the other week she had been found wandering more than a mile out of her proper course. He would take this opportunity of warning his satellite that his Astronomer-

142

Royal was watching her very carefully, and, if any further irregularities were reported, she would be liable to arrest as a vagrant and a vagabond. (Sensation.)

Continuing, the Earth deplored the eclipse policy pursued by the Satellites' Union, but said that as the law now stood they had no option but to submit. During the total eclipse in New York in 1925 a considerable sum was expended in municipal lighting, so that in that area at least the aim of the Union to achieve total darkness was frustrated. He hoped that the meeting would approve the proposals to limit the activities of the satellites to their legitimate functions.

The Moon, in a becoming apogee, seemed more than usually pale. She had heard with profound amazement the speech to which they had just listened. To approach the matter from the angle of the previous speaker must, she thought, mark the terminator of all mutual attraction, and their relations must henceforth enter on the penumbral stage. Throughout she had acted in the best interests of her principal, and thought that the attractive spectacles provided by the eclipses were welcomed by humanity at large. There was no question at all of intimidation. The times and places of each eclipse programme were widely advertised, and in all cases the quality and punctuality of the performance were guaranteed.

This work, the speaker emphasized, was a mere by-product of the ordinary activities of the Union to which she belonged. They laboured under grave secular inequalities, and she thought that their services were entitled to more adequate recognition. As they all knew, the chief work of the satellites lay in the maintenance of the tidal services and in night illumination. Nobody could say that their duties were not conscientiously performed. But as members of the Solar System they would insist on their rights, and the Satellites' Union would ever seek to preserve these.

Proceeding, the Moon remarked that the Einstein improvements in the Law of Gravitation did not make for simplicity, and under the new regulations it had become increasingly difficult to work to a schedule. (Hear, hear.) She fully expected that other planets would be found even further

143

out of their reckoning than herself. (Disorder.) She would name no names. (Uproar, during which the Moon beamed gibbously at Mercury. The secretary, whose eccentricity is well known, made a rapid transit amid loud laughter.)

In conclusion, her Union stood for the freedom of the skies. For herself, she had always done her best and would appeal to their sense of fair play. (Applause.)

The planet Jupiter said that, in a lengthy experience with a large family of satellites, he had always found that Kepler's policy of equal areas for equal times had paid. Any discrepancies of long period should be referred to stellar aberration. With him, eclipses were a daily occurrence and had long ago lost any interest. While his own axis was inclined to favour the proposals, he could not but regard them as unnecessary. His relations with his satellites had always been most amicable.

The aged planet Saturn said he was glad to find himself in conjunction with the previous speaker. He wished to contradict a rumour that he was in the habit of devouring his children. The rumour, he believed, originated with Galileo, but was disseminated by the Satellites' Union. Like his honourable friend, he had a large family by whom he had always striven to do his duty. He was proceeding to refer with some longitude to the obliquity of the modern satellite when he was recalled sharply to the Plane of the Ecliptic.

Venus, who was lustrous in aphelion with a crescent tiara of absolute magnitude, said she thought the planets were altogether too hard on the dear satellites. For herself she would gladly help them all she could.

Mars, speaking from Opposition, thought that too much attention should not be given to the words of the last speaker. She was a single woman with no responsibilities. (Cries of 'Shame.') He was strongly in favour of restricting the motions of satellites. His own two, he declared, made him feel positively giddy. His youngest in particular was accustomed to 'make rings round him' three times as quickly as he could turn himself—no mean anomaly, he added, for the God of War.

Considerable perturbations now became apparent among the satellites present, and one or two planets in perihelion left the Ecliptic at the Descending Node. On the Chairman's attempting to compute an ephemeris, scenes of meteoric confusion supervened, and the meeting thereupon terminated in a planetary nebula.

Astronomical abberation

From *After All* by Clarence Day (New York: Alfred A Knopf) 1936, p 57.

While current attention is focused on some of the moons of Jupiter that are whipping around backwards, Clarence Day had earlier made a general comment on this problem:

Phoebe, Phoebe, whirling high
In our neatly-plotted sky,
Listen, Phoebe, to my lay:
Won't you whirl the other way?

All the other stars are good
And revolve the way they should.
You alone, of that bright throng,
Will persist in going wrong.

Never mind what God has said—
We have made a Law instead.
Have you never heard of this
neb-u-lar Hy-poth-e-sis?

It prescribes, in terms exact,
Just how every star should act.
Tells each little satellite
Where to go and whirl at night.

Disobedience incurs
Anger of astronomers,
Who—you musn't think it odd —
Are more finicky than God.

So, my dear, you'd better change,
Really, we can't rearrange
Every chart from Mars to Hebe
Just to fit a chit like Phoebe.

145

Creation of the Universe: a modest proposal

From *Contemporary Astronomy* 2nd edition, by Jay M Pasachoff (Saunders College Publishing) 1981. An earlier version appeared in the *Journal of Irreproducible Results*.

HOPKINS OBSERVATORY, WILLIAMS COLLEGE, WILLIAMSTOWN, MA. 01267

To: Universal Creation Foundation

Request for supplement to UCF Grant #000-00-00000-001 'Creation of the Universe'

This report is intended only for the internal uses of the contractor.

Period: PRESENT TO LAST JUDGMENT
Principal Creator: CREATOR
Proposal Writers and Contract Monitors: JAY M PASACHOFF AND SPENCER R WEART

BACKGROUND

Under a previous grant (UCF Grant #000-00-00000-001), the Universe was created. It was expected that this project would have lasting benefits and considerable spinoffs, and this has indeed been the case. Darkness and light, good and evil, and Swiss Army knives were only a few of the useful concepts developed in the course of the Creation. It was estimated that the project would be completed within four days (not including a mandated Day of Rest, with full pay), and the 50% overrun on this estimate is entirely reasonable, given the unusual difficulties encountered. Infinite funding for this project was requested from the Foundation and granted. Unfortunately, this has not proved sufficient. Certain faults in the original creation have become apparent, which it will be necessary to correct by means of miracles. Let it not be said, however, that we are merely correcting past errors; the final state of the Universe, if this supplemental request is granted, will have many useful features not included in the original proposal.

PROGRESS TO DATE

Interim progress reports have already been submitted ('The Bible,' 'The Koran,' 'The Handbook of Chemistry and

Physics,' etc). The millennial report is currently in preparation, and a variety of publishers for the text (tentatively entitled 'Oh, Genesis!') will be created. The Gideon Society has applied for the distribution rights. Full credit will be given to the Foundation.

Materials for the Universe and for the Creation of Man were created out of the Void at no charge to the grant. A substantial saving was generated when it was found that materials for Woman could be created out of Man, since the establishment of Anti-Vivisection Societies was held until Phase Three. Given the limitations of current eschatological technology, it can scarcely be denied that the Contractor has done his work at a most reasonable price.

SUPPLEMENT

We cannot overlook a certain tone of dissatisfaction with the Creation which has been expressed by the Foundation, not to mention by certain of the Created. Let us state outright that this was to be expected, in view of the completely unprecedented nature of the project. The need for a supplement is to be ascribed solely to inflation (not to be confused with expansion of the Universe, which was anticipated). Union requests for the accrual of Days of Rest at the rate of one additional Day per week per millenium ($Dw^{-1}m^{-1}$) must also be met. Concerning the problem of Sin, we can assure you that extensive experimentation is under way. Considerable experience is being accumulated and we expect a breakthrough before long. When we are satiated with Sin, we shall go on to consider Universal Peace.

We cannot deny—in view of the cleverness of the Foundation's auditors—that the bulk of the supplemental funds will go to pay off old bills. Nevertheless we do not anticipate the need for future budget requests, barring unforeseen circumstances. If this project is continued successfully, additional Universe-anti-Universe pairs can be created without increasing the baryon number, and we would keep them out of the light cone of the Original. By the simple grant of an additional Infinity of funds (and note that this proposal is merely for Aleph Null), the officers of the Foundation will be able to

147

present their Board of Directors with the accomplished Creation of one or more successful Universes, instead of the current incomplete one.

We will attempt to minimize the additional delays that may temporarily exist during the changeover from fossil fuels to fusion for some minor locations in the Universe. For the time being, elements with odd masses (hydrogen, lithium, etc) will be created only on odd days of the month and those with even masses (helium, beryllium, etc) on the even days, except, of course, for Sundays. The 31st of each month will be devoted to creation of trans-uranic elements. The Universe has been depleted of deuterium; new creation of this will take place only on February 29th in leap years.

PROSPECTIVE BUDGET

Remedial miracles on fish of the sea	∞
Remedial miracles on fowl of the air	∞
Creeping things that creep upon the earth, etc	∞
Hydrogen	n/c (created)
Heavier elements	n/c (nucleosynthesis)
(Note: The carbon will be reclaimed and ecologically recycled.)	
Mountains (Sinai, Ararat, Palomar)	∞
Extra quasars, neutron stars	∞
Black holes (non-return containers)	∞
Miscellaneous, secretarial, office supplies, etc	∞
Telephone installation (Princess model, white, one-time charge, tax included)	$16·50

SALARIES

Creator (1/4 time)	at His own expense
Archangels	
Gabriel	1 trumpet (Phase 5)
Beelzebub	miscellaneous extra brimstone (low sulphur)
Others	assorted halos
Prophets	
Moses	stone tablets (to replace breakage)

| Geniuses | finite |

NB Due to the Foundation's regulations and changes in Exchange Rates, we have not yet been able to reimburse Euclid (drachmae), Leonardo (lire), Newton (pounds sterling), Descartes (francs), or Reggie Jackson (dollars). Future geniuses will be remunerated indirectly via the Alfred Nobel Foundation.

| Graduate students (2 at 2/5 time) | reflected glory |

MONITORING EQUIPMENT AND MISC.

1 5-meter telescope (maintenance)	finite
Misc. other instruments, particle accelerators, etc	large but finite
Travel to meetings	∞
Pollution control equipment	+40%
Total	$\infty + 40\% = \infty$
Overhead (51·97)	∞
Total funds requested	∞

Starting date requested:

Immediate. Pending receipt of supplemental funds, layoffs are anticipated to reach the 19·5% level in the ranks of angels this quarter.

Creation

Letter from Emil Herzog, quoted by R A Oriti (1980) *Griffith Observer* 44 (8) 20.

The astronomer Fritz Zwicky (1898–1974) was once engaged in a conversation with a fundamentalist minister regarding the origin of the universe. The minister said that the universe began when God said, 'Let there be light.' Zwicky, with a thoughtful expression, replied that he could 'buy' this explanation only if God had said 'Let there be light and a magnetic field.'

Stressing the astronauts

[In earthy language Tom Wolfe tells what he learned through interviews with astronauts about their intensely competitive, often dehumanizing training.]

From *The Right Stuff* by Tom Wolfe (New York: Farrar Straus Giroux) 1979, pp 96–99.

In one human test chamber, the pressure was reduced until an altitude of 65 000 feet was simulated. It made one feel as if his entire body were being squeezed by thongs, and he had to force his breath out in order to bring new oxygen into his lungs. Part of the stress was in the fact that they didn't tell him how long he had to stay there. They put each man in a small, pitch-black, windowless, soundproofed room—a 'sensory deprivation chamber'—and locked the door, again without telling him how long he would have to stay there. It turned out to be three hours. They strapped each man into a huge human milkshake apparatus that vibrated the body at tremendous amplitudes and bombarded it with high-energy sound, some of it at excruciating frequencies. They put each man at the console of a machine called 'the idiot box.' It was like a simulator or a trainer. There were fourteen different signals that the candidate was supposed to respond to in different ways by pressing buttons or throwing switches; but the lights began lighting up so fast no human being could possibly keep up with them. This appeared to be not only a test of reaction times but of perseverance or ability to cope with frustration.

No, there was nothing wrong with tests of this sort. Nevertheless, the atmosphere around them was a bit . . . *off*. Psychiatrists were running the show at Wright-Patterson. Every inch of the way there were psychiatrists and psychologists standing over you taking notes and giving you little jot'em'n'dot'em tests. Before they put you in the Human Milkshake, some functionary in a white smock would present you with a series of numbered dots on a piece of paper on a clipboard and you were supposed to take a pencil and connect the dots so that the numbers beside them added up to certain sums. Then when you got out of the machine, the White Smock character would give you the same test again, presumably to see if the physical experience had impaired your ability to calculate. And that was all right, too. But they also had people staring at the candidate the whole time and taking notes. They took notes in little spiral notebooks. Every gesture you made, every tic, twitch, smile, stare, frown, every time you rubbed your nose—there was some White Smock standing by jotting it down in a notebook.

Every night the boys got together in the Bachelor Officers Quarters and regaled each other with stories of how they had lied their heads off or otherwise diligently subverted the inquiries of the shrinks.

In one test the interviewer gave each candidate a blank sheet of paper and asked him to study it and describe what he saw in it. There was no one right response in this sort of test, because it was designed to force the candidate to free-associate in order to see where his mind wandered. The test-wise pilot knew that the main thing was to stay on dry land and not go swimming. As they described with some relish later on in the BOQ, quite a few studied the sheet of paper and then looked the interviewer in the eye and said 'All I see is a blank sheet of paper.' This was not a 'correct' answer, since the shrinks probably made note of 'inhibited imaginative capacity' or some goddamned thing, but neither did it get you in trouble. One man said 'I see a field of snow.' Well, you might get away with that, as long as you didn't go any further . . . as long as you did not thereupon start ruminating about freezing to death or getting lost in the snow and running into bears or something of that sort. But Conrad . . . well, the man is sitting across the table from Conrad and gives him the sheet of paper and asks him to study it and tell him what he sees. Conrad stares at the piece of paper and then looks up at the man and says in a wary tone, as if he fears a trick: 'But it's upside down.'

This so startles the man, he actually leans across the table and looks at this absolutely blank sheet of paper to see if it's true—and only after he is draped across the table does he realize that he has been had. He looks at Conrad and smiles a smile of about 33 degrees Fahrenheit.

On the spot?

The biggest solar flare for six years reached its peak today over the American city of Boulder, Colorado, the Institute of Telecommunications Sciences reported.

The Daily Telegraph 6 September 1966

Confirmation of the discovery of a black hole in the remote hinterland of Nova Scotia. The absence of a distance on the sign is consistent with the strongly non-Euclidean nature of spacetime near a black hole (Roy L Bishop).

Cocking a Snootnik at Sputnik

[Lines written at the time of the launch of Sputnik I, by a harassed astronomer.]

Contributed by the author, David S Evans

Twinkle, twinkle, little Sput
How you drive me off my nut,
Day and night the people call
They want to know why you don't fall,
At what hour you'll next appear,
If you'll still be here next year,
If it's true you have on board
A teleset which can record
All that goes on here below,
Ring the Kremlin, let them know.
If they persist in troubling me
On their screens they soon will see
Vulgar gestures made by me.

Still in the galaxy

Astronomers discover ethyl alcohol in interstellar space—News item

JEANNE HOPKINS

From *Isaac Asimov's Science Fiction Magazine* (September 1979) p 22.

There's a beat among stars in celestial bars
    That eluded astronomers' eyes.
They had long been aware of the ferment Out There.
    But they couldn't discover the prize.

Simultaneous passes with dishes and glasses
    Then sent their morale through the roof.
They were simply amazed how their spirits were raised,
    And they quickly established the proof.

Though research may be slowed by their staggering load,
    They may find, if they give it a shot,
The sobering news reinforcing the views
    That the cosmos is running to pot.

A damsel in distress

JOHN SCOFFERN

From *Chemistry No Mystery* (London) quoted in *Humour and Humanism in Chemistry* by John Read (London: G Bell & Sons) 1947.

'Hydrosulphuric acid gas enters into the composition of certain mineral waters; Harrowgate water, for instance, contains a large quantity of it; and connected with this subject, I have an anecdote to relate to you.

It was a practice with those ladies who were particularly ambitious of possessing a white skin, to daub themselves with a preparation of the metal bismuth, which is one of those that sulphuretted hydrogen blackens. Now it is represented on creditable authority, that a lady made beautifully white by this preparation took a bath in the Harrowgate waters, when her fair skin changed in an instant to the most jetty black. Uttering a shriek, she is reported to have swooned; and her attendants, on viewing the extraordinary change, almost swooned too, but their fears in some measure subsided on observing that the blackness of the skin could be removed by soap and water.'

A chemist was once defined as a poet who has taken the wrong turn. It is fitting, therefore, that a chemist should endeavour to pay a tribute in verse to this fair victim of a lack of chemical knowledge:

THE SURPRIS'D LADY
We do not share her great surprise,
We know the law that underlies
This lady's change of hue:
For what she thought was H_2O
Contained some H_2S, *and so*
She came down in Group II.

The lamentable incident finds a concise interpretation in the following scheme, which for once literally deserves to be called a personal equation:

$$2\ Bi(OH)_2NO_3 + 3H_2S = Bi_2S_3 + 2HNO_3 + 4H_2O.$$
Pearl-white   Black ppt
                                (Group II)

The horrid dangers that environ the milk-white bismuth-using siren are so varied, unsuspected and insidious, as to suggest that the use of this cosmetic should be limited by law to Honours graduates in chemistry. For observe: if the user bathes in the waters of Harrowgate, she turns black, in accordance with the requirements of the above equation; if she sits too close to a coal-fire in the gloaming, a similar fate may overtake her at a critical moment; if she uses the preparation too often, her skin will become rough and red; if she leaves it on too long, she will slowly turn yellow; and if she swallows it by accident, she will develop *methaemoglobinuria*. The replacement of this menace by such innocent materials as zinc white and starch takes not the least place among the services which chemistry has rendered to mankind.

Chemical history is shaped by a maker of wine bottles

From *Chemistry International* **1** 32 (1980).

Probably no unit is more widely used in chemistry than the litre, yet relatively little is known about the eighteenth-century maker of wine bottles, Claude Émile Jean-Baptiste Litre. He first proposed the system of specifying volumes in terms of the mass of liquid a container would hold. He was the pre-eminent manufacturer of chemical glassware of his day— the first to produce precision graduated cylinders and the inventor of the burette. His graduated cylinders varied in internal diameter by less than 0.1% of their length, and were graduated into hundredths—or even thousandths—of their volume.

Unfortunately for historians, Litre did not keep a journal of his work or his personal life, and correspondence exchanged with his close friend Celsius has been lost. His fine glassware has not survived. A set of graduated cylinders donated to the Royal Society of England in 1765 were destroyed 47 years later during the preparation of nitrogen trichloride by Sir Humphrey Davy. Most details of Litre's life were inferred from the general literature of the period. This information reveals that Litre—unlike most chemists of his generation— enjoyed prosperity, good health and recognition during his lifetime. He was hard-working, but abstained from excesses, and had a placid composure.

But then, Litre was not really a chemist. He was a manufacturer of wine bottles, as was his father, grandfather, and great-grandfather. His family's wine bottles were well-known in the Bordeaux wine industry since the 1620s. No doubt, this family tradition of glass-working was a major influence when Litre began his scientific training at age 16. So too was his meeting with Anders Celsius four years later during a geological expedition to the Swedish Lapland.

Celsius served as a liaison between the French expedition and Swedish officials. A professor of astronomy at the University of Uppsala, Celsius was preoccupied with precise measurement. The friendship between Litre and Celsius introduced the final element that was to prepare Litre for the most important contribution of his life.

In 1763 at the age of 47, Litre prepared his major written work, *Etudes Volumetriques*. He chose a standard volume very close to the *flacon royal*, introduced by King Henri IV in 1595 to standardize taxation of wine. However, Litre recognized that this unit was arbitrary, and suggested that volumes be specified in terms of the mass of standard liquid that a container would hold. He suggested mercury. Litre's proposal for a rationalized system of units materialized 15 years after his death, when a commission was appointed to formulate such a system, headed by the mathematician Lagrange. From this commission was born the metric system in 1795. Litre's method of specifying volume was adopted, as was the standard liquid. The chemist Antoine de Fourcroy was apparently the first to suggest Litre's name for the unit of volume. Two hundred years after Litre's death, the litre was adopted as the fundamental unit of volume for the standardized international SI metric system (Systéme International).

This short sketch of the life of Claude Émile Jean-Baptiste Litre is based upon a most interesting three-page article written by Dr K A Woolner of the University of Waterloo (Canada). It appears in the 11th issue of the *Int'l Newsletter of Chemical Education*, published by the IUPAC Committee on Teaching of Chemistry. A copy of the newsletter may be obtained free-of-charge from Mr P D Gujral at the IUPAC secretariat.

To err is human . . .

From *Chemistry International* **3** 2 (1980).

Prof. M L McGlashan (of University College London) informed us of 'the misleading hoax about "Monsieur Litre" perpetrated in *Chemistry International* 1980, No. 1, p. 32'. Stating that the story is completely untrue and probably originated as a joke, he wrote to say 'It is almost unbelievable that so many otherwise reputable scientists should be so gullible as to swallow such an article, but I am sick and tired of pointing out to the convinced that it was a spoof. What it must have done to less sophisticated scientists!'

Among these less sophisticated scientists was the editor of this magazine. Failing to detect the intended humour of the story, he unwittingly published it as a legitimate article.

Considerable efforts are made to ensure that information presented in *Chemistry International* is accurate. However, unless there is cause for suspicion, it is assumed that communications from established scientists are sincere. During routine checking of this article, no one informed the editor that the biography of M. Litre was a spoof. The editor extends his apologies for confusion which arose from publication of this article.

A chemical hoax

Contributed by Robert M Joyce.

In 1828 the German chemist F Wöhler synthesized a natural product, urea, and destroyed the tenet that a 'life force' had to be involved in the making of natural products. The next 15 years saw a flurry of argument over the constitution and structures of organic compounds. The prominent Swiss chemist J J Berzelius advanced the doctrine of dualism, which held that every compound was formed by the union of two radicals, single or compound, that had opposite electrical polarities. If an electronegative radical (e.g. chlorine) could take the place of an electropositive one (e.g. hydrogen), in one of the two parts of a compound, the nature and polarity of that part would be drastically altered, and the product of such a substitution therefore could not be of the same chemical type as the original compound.

The other school, headed by the French chemists J A B Dumas and A Laurent, argued that the doctrine of dualism could not be reconciled with experimental observations. For example, chlorination of acetic acid gave trichloroacetic acid, which had properties not dissimilar to those of the parent acetic acid.

This attack on the 'establishment', which included Wöhler and the influential J von Liebig, was not taken lightly. The following jest pokes fun at the idea of conservation of chemical type in substitution by describing a series of chlorinations of manganese acetate in which the various constituents are replaced by chlorine, finally giving a product that is entirely chlorine, but with the properties of the original manganese acetate. Wöhler penned it as a letter to Berzelius, using the French language to needle Dumas and Laurent. A copy found its way to Liebig who, as editor of *Liebig's Annalen der Chemie und Pharmazie*, decided to publish it, ascribing it to the pseudonymous S C H Windler ('Schwindler'). Wöhler protested, in vain, that the joke would be lost on those who did not read French, and that at the least, the pseudonym should be of French origin, for example, Ch Arlatan.

It should be noted that, in 1840, there was not general agreement on relative atomic weights of elements. Thus the first formula in the article was derived by replacement of 6 hydrogens by 6 chlorines and of 1 oxygen by 2 chlorines; the second was derived by replacement of manganese by 2 chlorines; the third was derived by replacement of 3 oxygens by 6 chlorines and of 4 carbons by 8 chlorines.

Over the following few years, accumulating experimental observations on substitution of hydrogen in various compounds by chlorine, bromine, iodine, nitro, etc became impossible to reconcile with the theory of dualism, and soon Berzelius was left alone in futile efforts to modify his theory to account for the facts. The theory of organic chemical structure was coming of age.

158

Brief communication to J(ustus) L(iebig)

Paris, 1 March 1840

Sir:

I hasten to describe to you one of the most astonishing findings in organic chemistry. I have verified the Theory of Substitution in a most remarkable and unexpected way. Only now can one appreciate the great value of this ingenious theory and foresee the important discoveries that it now promises. The discovery of chloroacetic acid and the immutability of types in chlorine compounds derived from ethyl ether and ethyl chloride have led me to carry out the experiments that I now describe. I have passed a stream of chlorine through a solution of manganese acetate that was exposed to sunlight. After 24 hours I found in the liquid a mass of crystals of a yellow-violet salt. The solution contained only the same salt and hydrochloric acid. Analysis of the salt showed it to be the chloroacetate of manganese protoxide. Thus far nothing surprising, only simple substitution of the hydrogen of acetate by the same number of chlorines, which was already known from the elegant research on chloroacetic acid. Heating this salt at 110°[C] in a stream of dry chlorine liberated oxygen and converted the salt to a new gold-yellow compound, analysis of which indicated the composition $MnCl_2$ + $C_4Cl_6O_3$. It appears that substitution of oxygen of the base by chlorine had occurred, which has been observed in many situations. The new compound dissolved in pure chloral on heating; I used this solvent, which is not affected by chlorine, to continue my study of chlorination. I passed dry chlorine into this solution for 4 hours while keeping it almost at its boiling point. During this period there was deposited continuously a white substance that careful examination revealed to be manganese protochloride. I cooled the solution when no more precipitate appeared and obtained a third substance as small needle-shaped yellow-green crystals. These analysed as $C_4Cl_{10}O_3$, corresponding to manganese acetate in which all of the hydrogen and manganese oxide had been replaced by chlorine. This formula should be written as $Cl_2Cl_2 + C_4H_6O_3$. There are thus 6 atoms of chlorine in the acid, the other 4 chlorine atoms representing manganese oxide. Like hydrogen, manganese and oxygen can be replaced by chlorine; this substitution does not appear surprising.

But this was not the end of this remarkable series of substitutions. On bubbling chlorine through a water solution of this new material, carbon dioxide was evolved and, on cooling the solution to 2°[C] there precipitated a mass of yellow platelike crystals that strongly resembled chlorine hydrate. They contained only chlorine and water. However, on determining the vapour density of the compound I found that it was comprised of 24 atoms of chlorine and 1 molecule of water. Thus there was achieved the complete substitution of all of the elements of manganese acetate. The formula of this compound should therefore be written as $Cl_2Cl_2 + Cl_6Cl_6Cl_6 + H_2O$. Although I had known that in the decolourizing action of chlorine there was replacement of hydrogen by chlorine, and that materials that are now bleached in England according to the substitution rules conserved their types,† I believe nevertheless that the substitution of carbon by chlorine, atom for atom, is my discovery. Please note this in your journal.

S C H Windler

† I have just learned that there are now in London shops materials made of chlorine filaments, that are widely used in hospitals and are preferred over everything else for nightcaps, underclothing, etc.

REFERENCES
Ihde A J 1964 *The Development of Modern Chemistry* (New York: Harper & Row) pp 191–198.
Pattison M M & Muir M A 1906 *History of Chemical Theories and Laws* (New York: John Wiley) pp 256–263.
Read J 1947 *Humour and Humanism in Chemistry* (London: G Bell & Sons) pp 214–217.

Early molecular biology: a hypothesis of the benzene ring in terms of
Macacus cynocephalus. Taken from a spoof issue of *Berichte der Deutschen
Chemischen Gesellschaft* **19** 3517–3568 (1886) whose title page read:

Berichte
der
Durstigen
Chemischen Gesellschaft
Unerhörter Jahrgang
No. 20.

When Albert Szent-Györgyi submitted his discovery of Vita-
min C to *Nature* he named it 'ignose' as it was a sugar of
unknown composition. However this term was rejected on the
grounds of flippancy and so Szent-Györgyi sent back his
paper with the compound retitled 'godnose'.

Professionalism—to the end

Communicated by
Julian R
Goldsmith.

Victor Moritz Goldschmidt, the renowned geochemist, was in Norway prior to the German take-over, and carried a container of potassium cyanide in the event of his arrest. Paul Rosbaud (who had been active with the underground during the war, and who spent many years before his death with Pergamon Press) knew Goldschmidt very well, and told me that a Norwegian colleague once asked Goldschmidt for some of the poison in case he too were 'taken'. Goldschmidt answered, 'Herr Professor So-and-so, poison is for chemists. You are a professor of mechanical engineering, and should use the rope.'

The purist

From *Verses from
1929 On* by Ogden
Nash (Boston:
Little and Co.)
1959.

I give you now Professor Twist,
A conscientious scientist.
Trustees exclaimed 'He never bungles!'
And sent him off to distant jungles.
Camped on a tropic riverside,
One day he missed his loving bride.
She had, the guide informed him later,
Been eaten by an alligator.
Professor Twist could not but smile.
'You mean' he said 'a crocodile.'

The ant

The ant has made himself illustrious
Through constant industry industrious.
So what?
Would you be calm and placid
If you were full of formic acid?

Fellowshipmanship

A concise manual on how to be a specialist without studying

IAN ROSE

Condensed from
'Fellowshipmanship'
*The Canadian
Medical Association
Journal* **87** 1232–5
(8 December 1962)

We begin with the statement of the insignia of the College for Fellowshipmen: *'Qui s'excuse, s'accuse.'*

THE APPROACH TO GENERAL PRACTITIONERS

It is as well at the outset to choose the image one intends to present to general practitioners, [with] the intention of putting the GP at a psychological disadvantage. Whatever role is adopted, the following rules must always apply:

(1) The Fellowshipman must always appear to be fond of the general practitioner, in the same way as a good citizen is fond of his dog, or the Mountie of his horse.

(2) The Fellowshipman must never openly disagree with the general practitioner on matters pertaining to medicine because there are far more effective ways of demonstrating his error and because the general practitioner is sometimes right. (This must never be admitted openly.)

MANNERIZATION I

Some nondescript and totally meaningless mannerism should be carefully developed, such as a tut-tutting noise made with the tongue against the teeth, or a significant but ambiguous raising of the eyebrows. Such mannerisms are invaluable in circumstances where a categorical 'yes' or 'no' might lead one to commit oneself to a definite opinion.

It is generally wise to ask the general practitioner to outline the history before going to the patient, so that he has to recount it from memory without the use of his notes. If the history is presented in a sketchy manner the Fellowshipman should pursue it to great lengths. If, on the other hand, the history is presented efficiently and clearly, the Fellowshipman should yawn, shift from foot to foot, examine his nails and make it clear that he feels it all to be a waste of time. A number of history-devitalizing questions can be used to interrupt and break up the flow of the general practitioner's presentation. For example, 'Just a minute. Was there a gag reflex present at the time?' or 'I suppose the father wasn't a glass blower, was he?' or 'Is the patient red–green colour blind?'

THE PRESENTATION OF THE DIAGNOSIS

This is the crucial point of the consultation and the pivot of the Fellowshipman's activity. His whole approach should rise to a crescendo, the peak of which is the pronouncement of the diagnosis, and all that may follow should be diminuendo in an atmosphere of glory and adulation. The opening gambit in presenting the diagnosis should be calculated to intrigue and mystify. It can be of two kinds:

(a) The fellowshipman leans back, lights a cigarette or cigar, and then says 'Well, it isn't a Weir–Westergrand–Jones syndrome.' Any questioning of it can be passed off since the Fellowshipman has just said that this is not the diagnosis in any case.

(b) The second gambit is 'Obviousmisting'. The Fellowshipman begins by uttering some such remark as 'It is of course essential to note that this patient has normal vision.' Alternatively, 'It is not without significance that the liver is not enlarged.' This gives the general practitioner something very puzzling to consider during the rest of the discussion.

Circumstances will dictate the final method of presenting the ultimate diagnosis. Under no circumstances should a positive diagnosis be given. In fact, the more obvious it is the more it should be obscured. Thus, it is not permissible in the case of a patient who has consolidation of his right lower lobe, a fever of 104 °F and pneumococci in his sputum, to make a diagnosis of lobar pneumonia. In such a case, the double negative diagnosis may be used to some effect. Thus: 'It is impossible to say that this patient does not have pneumonia.' The Fellowshipman may then select a special investigation (irrelevant as it may be), turn to the general practitioner and ask him for the results of this procedure—with the implication that if this had been done previously there would be no need to disturb the Fellowshipman from his Thursday afternoon golf.

FELLOWSHIPMANSHIP AND THE SPECIALIST

It is essential to know, before meeting the specialist, his age and where he trained; and any papers he may have published should be studied. The names of one or two senior men of his medical school should be committed to memory.

The opening gambit in these instances is most important. Even if it is not possible to intimidate the specialist at this early stage, it is quite essential to place him in an inferior light to any others present. Any one of the following may be used:

FELLOWSHIPMAN: What medical school did you go to?
SPECIALIST: Upperwoldingham and Ham . . .
FELLOWSHIPMAN: Well, then, you must have known Stinky Carpenter! (Carpenter was the Professor of Medicine at the time the Specialist was a medical student.)

Another useful opening is: 'I think I read a paper you wrote in such and such a journal a few months ago.' The specialist will then eagerly supply the title at which the Fellowshipman looks a trifle embarrassed and immediately suggests they get on with the business at hand.

MANNERIZATION II

A negative mannerism should also be cultivated. As an example, we may cite that developed by Garth Pindermoss, one of our honoured graduates. Pindermoss would entice a fellow specialist into describing a recent operation he had performed. At a crucial point he would ask an apparently simple question, such as: 'Did you undersew the superior inverted flap?' Whether the specialist says he did or did not, Pindermoss would then respond with the full negative mannerism. This consists of a half closing of the eyes, a partial frown, a slight shaking of the head, and the sound produced by drawing the breath sharply through pursed lips. This whole movement was beautifully executed by Pindermoss and could generally be guaranteed to have a paralyzing effect on the other specialist. During the subsequent course of the description of the operation, it was never necessary for Pindermoss to re-execute the entire manoeuvre, but at any time he could engender a similar emotional catastrophe in his opponent by making use of any one of the separate parts, such as half-closing the eyes.

If the argument is progressing to the Fellowshipman's disadvantage, a number of moves are available. For example, the Fellowshipman first lulls the specialist into a false sense of security by seeming to be persuaded. This creates a favourable impression of impartiality and scientific open-mindedness. Then the following manoeuvre is executed:

FELLOWSHIPMAN: Very interesting. A new approach! It totally invalidates the fundamental work of Rosengrats and Gildenstern.
SPECIALIST: How do you mean?
FELLOWSHIPMAN: Well, it's obvious, isn't it? (Pause.) You don't mean to tell me that you are putting forward this theory without comparing it to their work?

At this point it is very simple to show that the specialist has never given adequate thought to his thesis. The faith of the audience in the specialist is completely shaken, not only in relation to the present argument but in relation to his general standard of reasoning.

GENERAL RULES

(1) Never commit a diagnosis.

(2) The most recent work is always unpublished.

(3) The most recent published work is always in German.

(4) General practitioners are never right but may on occasion not be wrong.

(5) All specialists (other than the Fellowshipman) are always wrong.

(6) The most important symptom is one not related to the patient's present illness.

(7) The most important sign is the one no one else has found.

(8) No general practitioner is ever permitted to hear an early diastolic murmur.

Outline of evolution as dimly recalled from college education

From *Winnowed Wisdom; A New Book of Humour* by Stephen Leacock (New York: Dodd, Mead & Co) 1926, pp 15–17.

We are all descended from monkeys. This descent, however, took place a long time ago and there is no shame in it now. It happened two or three thousand years ago and must have been after and not before the Trojan war.

We have to remember also that there are several kinds of monkeys. There is the ordinary monkey seen in the street with the hand organ (*communis monacus*), the baboon, the gibbon (not Edward,) the bright, merry, little chimpanzee, and the hairy ourang-outang with the long arms. Ours is probably the hairy ourang-outang.

But the monkey business is only part of it. At an earlier stage men were not even that. They probably began as worms. From that they worked up to being oysters; after that they were fish, then snakes, then birds, then flying squirrels, and at last monkeys.

The same kind of change passed over all the animals. All the animals are descended from one another. The horse is really a bird, and is the same animal as the crow. The differences between them are purely superficial. If a crow had two more feet and no feathers it would be a horse except for its size.

The whole of these changes were brought about by what is called the Survival of the Fittest. The crookedest snake outlived the others. Each creature had to adapt itself or bust. The giraffe lengthened its neck. The stork went in for long legs. The hedgehog developed prickles. The skunk struck out an independent line of its own. Hence the animals that we see about us —as the skunk, the toad, the octopus, and the canary—are a highly selected lot.

This wonderful theory was discovered by Charles Darwin. After a five-year voyage in the *Beagle* as a naturalist in the Southern Seas, Darwin returned to England and wrote a book called *Sartor Resartus* which definitely established the descent of mankind from the avoirdupois apes.

Struthiomimus

or

The danger of being too clever

J MAYNARD SMITH

These stanzas are reprinted in *The Pattern of Vertebrate Evolution* by L B Halstead (San Francisco: W H Freeman), 1969.

[*Professor Smith's poem about extinction concerns the small bipedal dinosaur* Struthiomimus, *a late one of the carnivorous dinosaurs with long fingers, by which some believe it could steal the eggs of fellow dinosaurs. This contribution was written for Professor J B S Haldane on the occasion of his sixtieth birthday; in the original version the penultimate line started 'Remember Prof.'*]

The Dinosaurs *or so we're told,*
Were far too imbecile to hold
Their own against mammalian brains;
Today not one of them remains.
There is another school of thought,
Which says they suffered from a sort
Of constipation from the loss
Of adequate supplies of moss.

But Science now can put before us
The reason true why Brontosaurus
Became extinct. In the Cretaceous
A beast incredibly sagacious
Lived and loved and ate its fill;
Long were its legs, and sharp its bill,
Cunning its hands, to steal the eggs
Of beasts as clumsy in the legs
As Proto- *and* Triceratops,
And run, like gangster from the cops,

To some safe vantage-point from which
It could enjoy its plunder rich.
Cleverer far than any fox
Or Stanley *in the witness box*
It was a VERY GREAT SUCCESS.
No egg was safe from it unless
Retained within its mother's womb,
And so the Reptiles met their doom.

The Dinosaurs were most put out
And bitterly complained about
The way their eggs, of giant size,
Were eaten up before their eyes,
Before they had a chance to hatch,
By a beast they couldn't catch.

This awful carnage could not last;
The age of Archosaurs *was past.*
They went as broody as a hen
When all her eggs are pinched by men.
Older they grew, and sadder yet,
But still no offspring could they get.
Until at last the fearful time, as
Yet unguessed by Struthiomimus
Arrived, when no more eggs were laid,
And then at last he was afraid.
He could not learn to climb with ease
To reach the birds' nests in the trees,
And though he followed round and round
Some funny furry things he found,
They never laid an egg not once,
It made him feel an awful dunce.
So, thin beyond all recognition,
He died at last of inanition.

MORAL
This story has a simple moral
With which the wise will hardly quarrel;
Remember that it scarcely ever
Pays to be too bloody clever.

J B S Haldane in reply to a theologian who had asked him what evolution told him of God—'All I can say is that the Almighty showed an inordinate fondness for beetles.'

The dinosaur

WILL CUPPY

From *How to Become Extinct* (New York: Farrar & Rinehart) 1941.

The brain of a Dinosaur was only about the size of a nut, and some think that is why they became extinct. That can't be the reason, though, for I know plenty of animals who get by with less.[1] The Stegosaurus had a much larger secondary brain, or ganglion, in the pelvic region, so his thoughts were not on a very high plane. He didn't care what happened above the hips.[2] The Age of Reptiles ended because it had gone on long enough and it was all a mistake in the first place. A better day was already dawning at the close of the Mesozoic Era. There were some little warm-blooded animals around which had been stealing and eating the eggs of the Dinosaurs, and they were gradually learning to steal other things, too.[3] Civilization was just around the corner.

[1] The one interesting fact about the Diplodocus is that the accent is on the second syllable.
[2] The animal mind was not perfected until the Pleistocene Period, when it developed the ability to worry.
[3] This kind of progress is called evolution.

R W Wood

From 'Potpourri and Gallimaufry' by Paul E Klopsteg, *Science* **140** 594 (1963).

Probably no scientist has had more stories told about him, and probably none has told more stories about himself than R W Wood. Some 15 years ago a book was published with the title, *Dr Wood*, by William Seagrave, who had written two earlier books: *Asylum* and *Voodoo*. The unkind comment was made at the time that this completed the trilogy. Many of the anecdotes in the book were told exactly as they had been related many times by Wood himself. They were accounts of unusual and mostly amusing events. One which Wood told me was not included.

He expressed the view that if he were remembered after his death it would be not for his accomplishments in physical optics, but for his authorship of a small book with board covers, published in 1907 with the title, *How to Tell the Birds from the Flowers*. Each page had whimsical sketches of a bird and a plant which resembled the bird, with doggerel verses pointing out the differences. Wood had done both the sketches and the rhymes and was proud of his authorship.

Sometime in the early 1920s he had a Japanese graduate student at Johns Hopkins. One day the student, with some diffidence, said 'I understand, Dr Wood, that you wrote book about how to tell birds from flowers.' Wood acknowledged having done this, and went on to say 'The little book has long been out of print, and if you should locate one in a bookshop, you would probably have to pay as much as ten dollars for it.' The student thanked him. A few days later he returned and said 'Dr Wood, you told me about your book the other day.' Wood replied 'Did you locate one?' 'Yes, Professor, I did.' 'You were lucky to find one. How much did you have to pay for it?' 'Twenty-five cent.' After a pause the student resumed, 'Professor Wood!' 'Yes, what is it?' 'I find two copy.'

The Pansy.   The Chim-pansy.

From *How to tell the Birds from the Flowers—and other Wood-cuts by R W Wood* (New York: Dover) 1959 edn.

Observe how Nature's necromancies
Have clearly painted on the Pansies,
These almost human counten-ances,
In yellow, blue and black nu-ances.
The face however seems to me
To be that of the Chim-pan-zee:
A fact that makes the gentle Pansy,
Appeal no longer to my fancy.

The dry-rot of our academic biology

WILLIAM MORTON WHEELER

Condensed from
Science **57** 61
(1923).

Our Society requires its retiring President to close the annual meeting with a discourse or sermon. Text and title having been selected, autopsychoanalysis, which, like prayer, is now one of my favourite diversions, revealed the fact that I was suffering from an acute, repressed desire to commit sabotage on our academic biology by hurling a monkey-wrench into its smug machinery. My mental condition is, no doubt, partly due to the disappointing spectacle of our accomplishments as more or less decayed campus biologists in increasing the number, enthusiasm and enterprise of our young naturalists. In our universities, apart from the students preparing to enter medicine, the number indulging in advanced and graduate courses in the science would probably shrink to zero if we failed to provide fellowships or to hold out to them at the end of a long pole that enhaloed bundle of hay, the doctor's degree.

Is it the fault of the students? Obviously not, for no country produces a greater and more sweetly docile mass of pedagogical cannon-fodder. It would seem, therefore, that the teaching of biology should not be entrusted to those whom Bismarck called the damned professors, or that there is something wrong with us who try to teach the science, or with the environment in which we carry out the business. I can not avoid the impression that the problem involves, in varying degrees, all three of these factors. If you can bear with me, after a day of strenuous attention to far worthier utterances, I shall first consider very briefly some of the disabilities, both material and personal under which we seem to be labouring, and in conclusion suggest what I believe might be an ameliorative if not a remedial plan of action.

Any one must be staggered by the difficulty of selecting the most appetizing, concentrated and nourishing food for the student just entering the academic cafeteria. The student's metabolism may require plain gruel and toast, but we often insist on filling him up with so many elaborate pastries and salads that we ruin his digestion and, what is a thousand times worse, his appetite.

We might regard it as a great handicap that we academic biologists, unlike our native woodchucks and muskrats, are compelled to be most active pedagogically during the annual glacial period, but our superior intelligence enables us to cope with that situation. Every autumn we lay in a few cans of soused dog-fish and pickled sea-cucumbers, coop up some guinea-pigs, earth-worms, cockroaches and fruit-flies, throw in a bag of beans and several bales of hay for the botanists— and we are prepared for the worst. We can now proceed to disentangle and unreel the infinite and ineffable complexity of organic reality. When the neophyte becomes nauseated with the mess we have provided we can encourage him, and incidentally heighten our own prestige, by telling him that he is learning to forecast and control the behaviour of organic nature, that he may shortly be able to make real live homunculi and regulate their mating habits, and all the pishpash with which, since the Neolithic age, other priests and other wizards have heartened their constituencies.

More important than the drawbacks I have hinted at are certain types of personality engaged in the business of teaching biology. Even if we concede that the damned professor is an extraordinary being because he has sufficient inertia to specialize for a life-time in a particular department of learning, we must admit that he will grow old. From a young Antaeus continually gaining fresh strength from each successive contact with concrete reality he will become a creature increasingly infatuated with generalizations, relationships, and hypothetical explanations, especially if they are of his own confection, and he will eventually drift into a stage in which words, formulae and imaginary entities become the very breath of his nostrils. He has been borne aloft to be slowly asphyxiated in the tenuous atmosphere of the unreal.

It seems to me that there are two periods when the young biologist is most susceptible to lethal infection by the Merulius spores that are continually being thrown off by his professors. One is his freshman year, when he should be stimulated to develop an enthusiastic, receptive attitude; the other his graduate year or years, when he may be expected to adopt an independent, adventurous and creative attitude towards his science. Of course, the treatment of advanced students is easy for any professor who will follow the excellent example of the late Professor Roland of Johns Hopkins. The story is told that he was once presented with a list of rules for teaching graduate

173

students and that he crossed out all the items and wrote beneath: 'Neglect them!' Despite this very convenient precept, many of us coddle our graduate students till the more impressionable of them develop the most sodden types of the father-complex.

Not only do many of us wear out our most valuable tissues converting the graduate students into mere vehicles of our own interests, prepossessions and specialties, but nearly all of us fail to excite in them that spirit of adventure which has in the past yielded such remarkable results in the development of our science. The finest example of this lack of vision is seen in the stolid indifference, especially in our eastern universities, to explorations and research in the remoter portions of our own country, in foreign lands and especially in the tropics.

We leave the advanced student and turn to the beginner. What portion of the science of life, that most concrete and most entrancing of all the sciences, ought we to administer to this suckling host of postadolescence? I answer: they should be fed during the first year on the simple oatmeal pap of ecology.

History shows that throughout the centuries, from Aristotle and Pliny to the present day, natural history constitutes the perennial root-stock or stolon of biological science and that it retains this character because it satisfies some of our most fundamental and vital interests in organisms as living individuals more or less like ourselves. By the latter part of that pedantic century, the eighteenth, such great reserves of observation and experimentation had accumulated in the stolon that it began to bud. Taxonomy, morphology, palaeontology, physiology began to shoot up, branch and differentiate, becoming independent specialties, developing their own methods, fictions, and hypotheses. In the middle of the nineteenth century, after the great voyages of exploration, the bud chorology, or geographical distribution appeared, and about the same time I G St Hilaire and Haeckel, wishing to emphasize the fundamental importance of adaptation, but mistaking the stolon for a bud, named it 'ethology' or 'ecology'. More recently another dear little bud, genetics, has come off, so promising, so self-conscious, but alas, so constricted at the base.

The distinction between the *internal* and *external* relations of plants and animals has of course always existed, but has only lately come into such prominence that it divides biologists more or less completely into two camps—on the one hand those who make it their aim to investigate the actions of the organism and its parts by the accepted methods of physics and chemistry; on the other, those who interest themselves rather in considering the place which each organism occupies, and the part which it plays in the economy of nature. Professor Haeckel some twenty years ago proposed to separate the study of organisms with reference to their place in nature under the designation of 'oecology'. The ecologist is justified in regarding the whole living world as an intricate congeries of consociations, ranging in complexity from at least two to a great many organisms. Even genetics may be regarded as a department of ecology. Hence the problem of adaptation is not foreign to this discipline. Moreover, since human societies are very intimate and elaborate biocoenoses of individuals of the same species, psychology, sociology, economics, anthropology, ethnology, history, ethics, jurisprudence, government, hygiene, medicine, etc, are essentially ecological, for their central problems are behaviouristic.

Although I have left our lusty young freshmen out in the cold during this long harangue, I have not forgotten them. I repeat: What ought we to give them? I do not believe that we should inform them with the first crack out of the box that they are animals and descended from ape-like ancestors. This must come as a severe shock to any young *Boobus americanus* who has never had an opportunity to make the acquaintance of really high-class apes. The freshman laboratory should be neither an animal morgue nor a herbarium, but a vivarium. Its teachers should be numerous, competent, enthusiastic, and young. Some means must be devised for taking the students into the field more frequently, since it is impossible to reproduce and study the more complex biocoenoses under artificial conditions.

There is another suggestion I should like to make, in order that the freshman course may be preserved from the dry-rot, which may invade even the most dynamic type of instruction,

Modelling the cat falling

HENRY PETROSKI

From *Creative Computing* (May–June 1978) p 34.

1. *The cat acquires a velocity*
   At thirty-two feet per second per second
   And everything begins to blur—the tail
   Is twirling, the cat is turning, the paws

   Are on the ground. Again the cat has turned
   From upside-down around to downside-down
   Without a wall to push on or a string
   To pull itself around on the way down

   What is the mechanism? What is the
   Solution to the problem of the cat,
   Released from rest and oriented up,
   Descending in a circle in a line

   Of gravity, the quickest thinker, down,
   To always land with four paws on the ground?

2. *In* Comptes Rendus, *in 1894,*
   The cat stop falls in photographs Monsieur
   Marey has taken with his camera.
   His explanation is a hopeless use

   Of words for pictures: everyone can see
   The cat superimposed upon himself:
   The first slow feet, and then the faster feet
   Prepared to meet the sidewalk half way down.

   Monsieur Marey contorts his vertebrae
   With words like torsion, opposition, tors
   And wraps his tail around his helix spine
   Once clockwise for a counterclockwise half

   Rotation of the animal. We see
   Two human hands still grasping for the cat.

3. *Herr Magnus (no computer) had to crank*
   The torso of the cat manually
   Through its manoeuvres. So he simplified
   His model for the numbers it would use.

Dr McDonald, physiologist,
Dealt with the phenomenological
Aspects. His cat was not an equation
That did its business neatly on the paper.

Professors Kane and Scher were fortunate.
A man on the moon supported them
And NASA sectioned cat cadavers so
The moments of inertia of the cat

That Kane and Scher did use were accurate
Enough to lend credence to their results

4. *A couple of cylinders is their cat,*
   Without a head, with negligible legs,
   Without a tail: a mechanical Manx.

   This cat possesses a Lagrangian,
   Potential and kinetic energy
   Confused in an expression for the cat

   Falling, the cat jumping, the cat at rest.
   The lithe Lagrangian, ready to be
   The cat in the clutches of gravity

   Submits to differentiation with
   Respect to time and with respect to speed
   To fall in a falling and revolving mode.

   Released now, the equations, upside down,
   Descend in the computer, and they turn.

REFERENCES

Marey M (1894) Des mouvements que certains animaux exécutent pour retomber sur leurs pieds, lorsqu'ils sont précipité d'un lieu élevé *C. R. Acad. Sci., Paris* **119** /14–7

Magnus R (1922) Wie sich die fallende Katze in der Luft umdreht *Archs. Neerl. Physiol.* **7** 218–22

McDonald D A (1955) How does a cat falling turn over? *St. Bart's Hosp. J.* **56** 254–8

Kane T R and Scher M P (1969) A dynamical explanation of the falling cat phenomenon *Int. J. Solids Structures* **5** 663–70

The speed of the deer fly

ROBERT L WEBER

In 1927 C H T Townsend reported on his studies of the anatomy of the *Cephenemyia* and his observations of their flight. 'On 12 000 foot summits in New Mexico,' he wrote, 'I have seen pass me at an incredible velocity what were quite certainly the males of *Cephenemyia*. I could barely distinguish that something had passed—only a brownish blur in the air of about the right size for these flies and without sense of form. As closely as I can estimate, their speed must have approximated 400 yards per second [818 mph].'

Soon afterward, an editorial in *The New York Times*, in commenting on a new seaplane record of something over 300 mph warned man not to be too boastful of his accomplishments, since the deer fly has a speed of 700 mph.

A two-page diagram in *Illustrated London News*, 1 January 1938, gave comparative speeds attained by man, and by animals, fish, etc. The male deer fly was credited with 818 mph, the female 614.

Irving Langmuir at the Research Laboratory of the General Electric Company was sceptical of these speeds claimed for the deer fly. Using a formula from ballistics, $R = \varrho l^2 v^2 f$, for a speed of 818 mph the flat head of the fly would experience a force R from air resistance of 1.0 newton, or an added pressure of more than half an atmosphere, probably enough to crush the fly. The power consumption needed to maintain this velocity would be 370 watts or about one-half horsepower—a good deal for a fly! Langmuir further calculated that to deliver 370 watts the fly must consume about 0.3 g of food per second, or 1.5 times his own weight of food each second.

To try to give numerical meaning to Townsend's 'barely distinguishable blur in the air,' Langmuir took a short piece of solder about 1 cm long and 0.5 cm diameter and tied it about its middle to one end of a light silk thread, holding the other end in his hand. With lengths of thread from 1 to 3 feet it was easily possible to swing the weight in a circle in a vertical plane at the rate of 3 to 5 rotations per second. In this way speeds from 13 to 61 mph were produced.

Observations in a room, with a brightly lighted white ceiling as background, showed that at 13 mph ($580\,\mathrm{cm\,s^{-1}}$) the 'fly' was merely a blur—the shape could not be seen, but it

180

could be recognized as a small object of about the correct size. At 26 mph ($1150\,\text{cm s}^{-1}$) the fly was barely visible as a moving object. At 43 mph ($1920\,\text{cm s}^{-1}$) it appeared as a very faint line and the direction of rotation could not be recognized. At 64 mph ($2000\,\text{cm s}^{-1}$) the moving object was wholly invisible.

Langmuir wrote 'In the Adirondacks I have been surprised by the swift flight of insects which I have been told were deer flies. These flies often strike one's bare skin with a very noticeable impact, far greater than that of any other insect I have met. However, if the speed were 800 mph and the fly on striking should be stopped within 1 cm it would come to rest in about 55×10^{-6} s, and during this time there would be a force of 1.4×10^3 newtons (310 pounds force). It is obvious that such a projectile would penetrate deeply into human flesh.

'The appearance of the moving lead weight on the thread at 25 mph seems to agree roughly with my recollection of the deer fly. A speed of 800 mph is utterly impossible.'

REFERENCES

Townsend, Charles H T (1927) On the Cephenemyia mechanism . . . *Journal New York Entomological Society* **35** 245–252.
Langmuir I (1938) The speed of the deer fly *Science* **87** 233–234.

A rich man interested in breeding horses commissioned three experts, a vet, an engineer, and a theoretical physicist, to find out their best properties. After a few years they reported their results. The vet had concluded from a genetical study that brown horses were the fastest while the engineer had found that thin legs were optimum for racing. The theoretical physicist did not give up his quest at the end of the period but merely asked for more time to study the question claiming that the case of the spherical horse was proving very interesting.

AARON KATCHALSKY

Designer genes: a new look in fish genetics

JOEL E COHEN

Two groups of scientists working independently in laboratories in New York and Texas have recently reported that all fish are the same size.

Although scientists elsewhere are reluctant to discuss this remarkable breakthrough until confirmatory experiments can be performed, the simultaneous announcement by widely separated research groups tends to confirm the discovery.

The apparent difference in size between a sardine and a tuna is an illusion, the scientists say. 'Some fish have a gene to make them look big. Some fish have a gene to make them look small. In fact, they are all the same size' said Dr Arnold Storey, of the Long Island Center for Ecobiology, in an interview. 'We have proved this by extracting the so-called shrimp gene from the DNA of sardines.' DNA is the chemical that carries the genetic information in mammalian cells. 'The shrimp gene makes the sardine look small. We cloned the gene by the techniques of recombinant DNA. We then injected many copies of the gene into tuna cells grown in a laboratory culture dish, after the original DNA from the tuna cells was removed. The removed DNA has been donated to an organ bank, where it is earning 18% a year.

'We implanted the hybrid sardine–tuna cells in a receptive aardvark which had been immunized by forced feedings of onions and sour cream. After three months of gestation in an air-conditioned apartment on the upper east side of Manhattan, the aardvark gave birth to a fully grown, perfectly formed tuna the size of a guppy. 'We consider this a remarkable achievement because the aardvark was male.'

At the Fisheries Institute of South Houston, workers performed the reverse experiment, transferring the gene that makes a tuna look big into a sardine. 'Animals control their apparent length or weight by mechanisms we call designer genes' a spokesman for the Texas group said, holding tightly onto a spoke. 'Without such genes, it would not be easy for a tuna to fit into a sardine can.'

According to members of both the New York and Texas groups, designer genes are destined to play an important role in explaining how genetic information controls morphogenesis, the process by which a single egg cell or ovum

182

develops into an intricately formed adult. 'Why do some ova grow into enormous sharks, while others end up goldfish?' Dr Storey asked. 'These designer genes must have something to do with it, or we wouldn't be studying them.'

The finding that fish vary in shape but not in size suggests that the mechanisms controlling morphogenesis are far subtler than anyone had previously suspected, the Texas researchers suggest. 'The intriguing question' Dr Shirley Wilde of Houston said 'is not what makes them all the same size, but what makes us see them as different. If we understood that, we would understand why Jonah was swallowed by a minnow disguised as a whale.'

The simultaneous announcement of the cloning and transfer of designer genes surprised both of the succesful groups.

'When we reviewed the grant application of the FISH group for the Science Foundation several years ago' Dr Storey said 'it seemed so ridiculous that we never expected them to get the money to do the research. So we decided to try it ourselves.' Dr Storey's research is funded by an industry association seeking proof that there is more fish in a 3 ounce can of tuna than there is in a 4 ounce can of sardines. In 1973, Dr Storey received the Outstanding Ichthyologist of Southeastern Long Island Award from a group of scientists in his employ.

Members of the Texas group learned of Dr Storey's efforts only when one of them was asked by a scientific journal to evaluate a research report Dr Storey had submitted for publication.

'There's a gene in every genius' said Dr Wilde 'I guess that's why we got the idea first.'

The reason life is probably extinct on other planets is that their scientists are a little more advanced than ours.

ANON

Curve fitting
Competitive interactions between neotropical pollinators and Africanized honey bees

From *Science* **201** 1030 (1978).

ABSTRACT

The Africanized honey bee, a hybrid of European and African honey bees, is thought to displace native pollinators. After experimental introduction of Africanized honey bee hives near flowers, stingless bees became less abundant or harvested less resource as visitations by Africanized bees increased. Shifts in resource use caused by colonizing Africanized honey bees may lead to population decline of Neotropical pollinators.

DAVID W ROUBIK

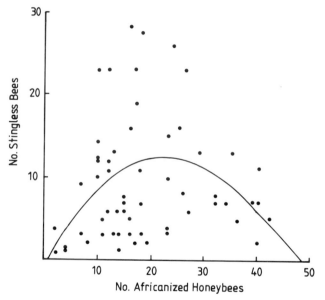

FIGURE 1. The relations of Africanized and stingless (meliponine) bee abundances on flowering *Melochia villosa*. The full line is a quadratic polynomial (given by $y = -0.516 + 1.08x - 0.023x^2$) which gave the best fit to the points.

From *Science* **202** 823 (1978).

The rather fanciful curve fitting of Roubik (figure 1) has prompted me to propose an alternative interpretation of his data (see figure 2).

ROBERT M HAZEN

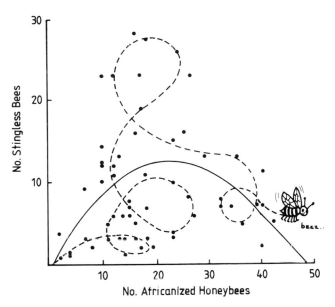

FIGURE 2.

In reply, Dr Roubik has commented 'I think Dr Hazen was right in being sceptical, but I do not think that it would justify disregarding the study or my conclusions. I thought that his letter to the editor was hilarious, but some of my colleagues did not. It seems to me that biologists are often obliged to take a different view of quantitative data from that of physical scientists. They have more or less set rules, while we must often try to discover nature's meaning. And there is a lot of slop in nature.'

Understanding atomic physics is child's play compared to understanding child's play.

ANON

Biorhythm: imitation of science

W S LYON, F F DYER and D C GARY

Condensed from
Chemistry **51** 5–7
(1978).

Rhythm: dance bands play it; the Pope approves it; Ethel Merman has it.

Now it seems there's a special brand of that delightful commodity which we all possess; the soothsayers, pseudo-scientists, and salesmen call it biorhythm. Maybe you've seen their tootings in the popular press, in advertisements, or on a TV talk show. They, too, have a special rhythm of their own, and it swings along to the tune of a jingling cash register and a jangling sales pitch that promises to predict your future and explain your past very scientifically.

According to biorhythm, every individual has three internal clocks or biorhythm cycles that begin at birth and keep precise time throughout the individual's life. The 'physical cycle' has a 23-day period, the 'emotional cycle' a 28-day period, and the 'intellectual cycle' a 33-day period. When the cycles are above the axis—as in the first half of their periods—the individual is physically strong, mentally awake, and morally straight, just as the Boy Scout Handbook says. When the cycles are in their second-half periods, strength, emotion, and intellect are in a weakened condition.

What happens when a cycle crosses the axis, as it does for example in figure 1 with the 28-day period on days 1, 14, and 28? These crossover days are times of greater vulnerability for an individual, and strange and terrible things may often happen. A double crossover is doubly dangerous, and on those relatively rare occurrences when all three cross over together, one might just as well echo the remark of promoter Joe Jacobs, 'I should have stood in bed.'

How did biorhythm originate? Martin Gardner gave the background of the movement some ten years ago in one of his Mathematical Games sections of *Scientific American* (July 1966). As Gardner pieced together the story, biorhythm was the brainchild of a German physician, Wilhelm Fliess, who developed his ideas in the period 1880–1900. Fliess was obsessed with the numbers 23 and 28 to which he attached mystical male and female properties. He was also the father of a highly unusual theory: that many physical illnesses were caused by, could be diagnosed by study of, and could be cured by proper treatment of—guess what? The nose!

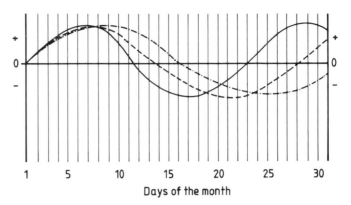

FIGURE I. Biorhythm chart of the three cycles showing crossover points at 11.5 days (physical), 14 days (emotional) and 16.5 days (intellectual).

When we learn that Sigmund Freud was a close friend, admirer, and even patient of Fliess, it comes as no surprise to learn that the nose and its mucous membrane were supposed to hold the answer to sexual hangups and genital problems in general. The friendship between Fliess and Freud broke under the strain of biorhythm: the former accused Freud of stealing his ideas and passing them to Hermann Swoboda, a patient of Freud who, Fliess claimed, broadcast them about without crediting Fliess. Fliess' fears were indeed solidly founded; today in the 'Sacred Writings' of biorhythm, Swoboda is invariably mentioned as one of the cult's founding fathers.

Beyond attracting a few followers in Germany during the early years of this century, biorhythm made little impact on the world. Beginning about ten years ago, however, the theory surfaced and began to be mentioned in what is politely called 'the media'. Somewhere along the way, a third cycle termed the 'intellectual' was added. Combining a bit of pseudoscience (the mathematical cycles) with a lot of mysticism (the prophetic power) biorhythm seemingly presents the best of two worlds in a single package. . . .

From the Industrial Safety and Applied Health Physics Division at Oak Ridge National Laboratory, we obtained a list of 112 lost-time accidents, the birthdate of the person who had the accident, and the date it happened. A computer program was written (to check predictions of biorhythm theory). The results are shown in table 1. As can be seen, there are no statistically significant events beyond those expected by random occurrences.

TABLE 1 Results from 112 major accidents at ORNL.

Cycle	Accidents falling on crossovers	Accidents expected by chance alone
Physical	18	15 ± 3
Emotional	6	8 ± 2
Intellectual	9	10 ± 2

These accidents were then divided into three classes: I, those where the victim was obviously at fault; II, indeterminate fault; and III, victim was innocent of blame. These were run through as three separate groups. Again, there was nothing statistically significant to bolster up the biorhythm theory.

Our conclusion therefore, is that biorhythm is bunk! We put it to the test using data from the workplace involving real accidents of the most serious nature. We could have taken a much larger sample containing thousands of minor accidents such as cut fingers, minor acid burns, bruised knees, etc, but we preferred to use severe accidents. Maybe you, as an interested reader, have some data you would like to test for biorhythm effects.

The original electrocardiograph built around 1911 by the Cambridge Instrument Co. Five people were needed to operate this machine—no wonder the victim looks worried (Cambridge Medical Instruments Ltd).

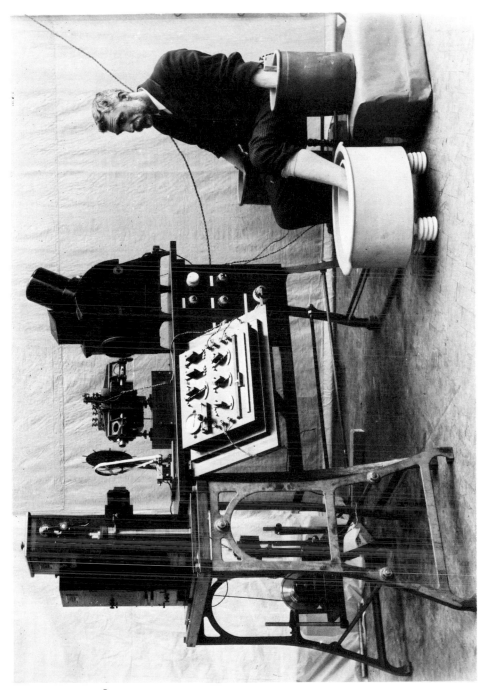

On being blinded with science

Being a ruthless enquiry into scientific methods, complete with 29 genuine references and a learned footnote

D E H JONES

Condensed from *New Scientist* (4 November 1966) p 1465.

An example of that rare and delightful phenomenon, the successful scientific hoax, appeared in *The Analyst*[1] as an account of work on the toxicology of ice, claiming to emanate from the Bureau of Chemical Investigation of the New York State police; and the story is that it was slipped past the editors of *The Analyst* by a disgruntled abstractor under notice. Such horseplay is very rare in the respectable world of modern scientific publication.

In the past, however, fake publications seem to have been more common, and those grand old men of nineteenth-century chemistry, Liebig and Wöhler, used their editorial positions to publish at least two disgraceful skits on the theories of their rivals. One of these (quoted in Bulloch's *History of Bacteriology*[2]) appeared anonymously in *Liebig's Annalen der Chemie* of 1839[3], and was intended to ridicule those who held that fermentation was a biological rather than a purely chemical process: it purported to describe in great detail how, with the aid of a super-microscope, the author had observed animalcules in the yeast eating up the sugar and excreting alcohol and carbon dioxide!

The other paper[4], published in the same journal by 'S C H Windler', was a similar heavy-handed attack, this time on the notion of 'substitution reactions' in organic chemistry, which were a novel concept at that time [see p 157]. This hoax reported, again in great detail, how by treatment of manganous acetate with chlorine the author had successively substituted more and more of the atoms of the original molecule by chlorine atoms, until he finally obtained a yellow solid that was entirely chlorine. The paper also commented that fabrics of spun chlorine (made by similar chlorine-treatment of normal textiles) were produced in England and were in great demand!

As late as 1886 a completely fake issue of the staid journal of the German Chemical Society was produced,[5] as a supplement continuing the pagination of the previous issue, and was printed and distributed through the usual channels [see p 161]. This, unlike the other two, had as far as I know no ulterior motive, but merely carried a number of humorous pseudo-scientific papers (much as does the modern *Journal of Irreproducible Results*). It is unfortunately rather rare today, the intervention of stern scientific librarians presumably having prevented its being bound into many of the standard sets of the journal, but a description of it is given by Sir John Read.[6]

There is a sense in which every scientific paper published partakes a little of the spirit of these travesties. The basically untidy and chaotic nature of research at the frontiers of knowledge is too far from the standard format of modern publication, with its straight-forward account of successful experiments and well-supported conclusions, to fit into it without some discrete recasting of the work. As R P Feynman, co-winner of the 1966 Nobel Prize for physics, remarked at the beginning of his excellent Nobel lecture:[7]

> We have a habit in writing articles published in scientific journals to make the work as finished as possible, to cover up all the tracks, to not worry about the blind alleys or describe how you had the wrong idea first, and so on. So there isn't any place to publish, in a dignified manner, what you actually did in order to get to do the work.

But the impression accordingly given of research as a solid, irresistible Sherlock-Holmesian advance into the unknown is almost certain to be misleading; and in the late 1950s a whimsical *Technical Glossary* appeared, purporting to give the real meaning behind the stylized phrases of research publications. Despite its exaggeration, this glossary said for the first time something so basic about the literature that it was rapidly copied, plagiarized, quoted, and ascribed to various sources,[8,9,10,11,12] and has now passed into the folklore of science; to my best knowledge the actual originator was C D Graham Jr,[13] of General Electric, and the glossary is perhaps most accessible in the splendid compilation *The Scientist Speculates*.[14] I cannot resist quoting some of the more enlightening entries:

Three of the samples were chosen for detailed study: The results on the others did not make sense and were ignored.

Correct to an order of magnitude: Wrong.

Typical results are shown: The best results are shown.

191

While it has not been possible to provide definite answers to these questions: The experiments didn't work out but I figured I could at least get a publication out of it.

A new compilation has been published by D Kritchevsky and J R Van der Wal.[15] *The Scientist Speculates* has another article in the same vein by one (I strongly suspect pseudonymous) J de Bloggins[16] in which the applied, or realist, scientist is defined as one who accepts in his own field the importance of the three fundamental problems, or their equivalents, enunciated by Griffiths for engineering: Dirt, and Noise, and Leaks. In the face of this unacknowledged trinity, not to mention the ubiquitous Bugs or Gremlins†, no wonder it is much more difficult than the bland accounts of successful experiments would imply to wrest an unambiguous reply from Nature, or to devise some piece of apparatus and make it work. In fact it is probable that about 80% of scientific experiments fail, in the sense that they do not give the information they were originally designed to elicit (which is why the Russian efforts to represent their space effort as having met from the beginning with total success are so unconvincing).

The same uncertainty holds good in the theoretical field. Deducing workable hypotheses from experimental data is a very uncertain business: except in mathematics (which is merely the art of saying the same thing in different words), every chain of reasoning rapidly loses conviction as you pursue it, and frequent experimental checks are needed to convince you of its continued accuracy. This is why it is so difficult to believe the immortal Sherlock Holmes sagas, in which the great detective ostensibly applies scientific methods with such astonishing success. We instinctively recognize the unworkability of Sherlock's guiding principle,[19,20] 'When all other contingencies fail, whatever remains, however improbable must be the truth' for in practice there are always too many possibilities. In fact I know of only one use of this

---

† I do not myself believe in Bugs or Gremlins, preferring to regard them as naive personifications of the Universal Law of Cussedness, expressed in Murphy's edict that, if a thing can go wrong, it will. For a theoretical understanding of this Law, and an account of the crucial Clark–Trimble experiment of 1935, see References[17,18].

principle (other than Sherlock's own contributions[21]) in the chemical literature. W N Lipscomb, writing on the structure of B_9H_{15} explained:[22] 'The boron arrangement was elucidated with the aid of the Remington Rand 1103 computer by the method of Holmes,' giving the above quotation from *The Adventure of the Bruce-Partington Plans* as reference.

The ability to be convinced by the identical results of repeated tests is such a marked characteristic of the scientific attitude that the unreproduced anomaly, even if carefully recorded, is useless for scientific purposes and has to be ascribed to the machinations of the Bugs. Scientific theories therefore, being built on data from which the irregularities have been removed, envisage a remarkably orderly picture of their subject. Even so, a good theory unifies so many different facts so well that it is difficult to resist the persuasion that it is not just a consistent explanation, but is really *true*—indeed, science would not be worth bothering with otherwise, for it is fundamentally motivated by the human need to understand what the world is all about.

Its outstanding success, however, is probably due less to any superhuman dedication on the part of its practitioners than to the emotional unimportance of its subject matter. Nobody cares enough one way or the other about (say) the existence of the quark to falsify the evidence or indulge in that casuistical special pleading 'the finding of bad reasons for things you believe for worse reasons' (as Aldous Huxley commented) so prevalent in philosophy, religion, and intellectual politics, which claim to dispense truth about emotionally more important facets of the World. Of course, where these latter have collided with science (as for example in Galileo's confrontation with the Catholic Church, the fundamentalist opposition to evolution, and Lysenko's political falsification of genetics) the situation has been more complex, but ultimately the strain of pretending that things are other than they observably are gets too intolerable. Great is the truth and it shall prevail!

Unfortunately, the scientist is not entirely in the ideally and idyllically simple situation of merely trying to observe the facts as accurately as he can, and then fitting them into the simplest theory. There are always distorting pressures, of which the most understandable is the desire to believe one's own theory rather than those of others: but obviously this effect is reciprocal and cancels out on the average. An amusing example is quoted by Pimentel and McClellan in the field of hydrogen bonding:[24] 'Batuev, attempting to marshall every argument in favour of his own frequency-modulation theory in preference to the pre-dissociation theory, observes that the pre-dissociation theory ignores the fundamental principles of Marxism–Leninism. Neither Marx nor Lenin appear to have published any work on the H bond.'

A more serious pressure results from the nature of accepted theory, which at any time covers, in principle, the whole range of observable phenomena with the possible exception of a few doubtful points. Cuvier remarked that the greatest service one can render to science is to make the ground clear before constructing anything there: but in fact this never occurs. Thus the tiny flaws in Newtonian physics were known twenty to forty years before they led to its replacement by relativistic concepts—but nobody abandoned Newtonian physics in the interval. An imperfect theory is always much better than no theory at all—an attitude reminiscent of the traditional moralist's fulmination against the iconoclastic young who 'want to knock things down without having anything to put in their place!'

As a consequence there is little to be gained by publishing results which conflict with current ideas without suggesting a satisfactory modification, for the obvious imputation is that they are plain wrong. None the less, such embarrassing results if substantiated are clearly of the highest importance, and it is impossible to guess how much is lost by the tendency to reject them. One example which came to light only by accident was the case of the benzocyclobutene system, first synthesized, as we now know, by Hans Finkelstein at Strasbourg in 1909 as part of his doctoral research. At that time it seems to have been held by chemists (for no particularly cogent reason) that benzocyclobutenes were impossible, so the finding, although it was complete and correct in all respects, was never published.

Accordingly the existence of this important ring-system remained unknown, and reasons for its impossibility were elaborated by theorists for another 46 years until a copy of Dr Finkelstein's thesis was accidentally discovered in 1955.[25,26]

Again, consider the case of Dr E J Saxl,[27] who had the misfortune to obtain with an electrically-charged torque-pendulum results which appear to demand major revisions of existing physical theory. He found that his apparatus did not behave symmetrically with respect to positive and negative charges, and was affected by astronomical phenomena (the changing seasons, eclipses, and so on). He did indeed publish the work, but with a reluctance obvious from his apologetic tone:

> The phenomena described here seem to go beyond what could be legitimately calculated from an extrapolation of present theory. The physicist hesitates to form a working hypothesis for such observations. Having accounted carefully, however, for all other influences, he is driven reluctantly to the conclusion that there may exist variations in g even if such cannot be noted with grounded, quasi-stationary instruments.

It is likely that almost every scientist in the course of his work encounters such weird effects, whether genuine or not, and it seems a pity that there is no safe way to attract attention to them. The early scientists like Galileo and Huygens had a technique for overcoming this problem: they published original but controversial findings in the form of anagrams or cryptograms, which they could later decipher to claim priority and credit if the discovery turned out to be well-founded, or quietly forget if it did not. While this is not perhaps practical today, there might be a place for Daedalus's *Journal of Odd Results*[20] in which such peculiarities could be anonymously recorded, but with the authors confidentially listed so that they could claim credit should it become possible to do so. The same organ could also perform an equivalent function in the field of theoretical speculation (a journal of '$\frac{1}{2}$-baked ideas'[29] has already been mooted, and led directly to the publication of *The Scientist Speculates*). It would thus act as a powerful scientific catalyst in which, cloaked in discreet anonymity, new ideas and facts could be brought together, and such *cris de coeur* expressed as:

'Satisfactory explanation required for bizarre phenomenon'
'Fascinating but naive hypothesis wishes to meet pertinent facts'
'Anxious experimentalist wishes to know whether a certain observation would be in conflict with current theory.'

Only one difficulty occurs to me. The spiritual descendant of Liebig and Wohler and the *Analyst* abstractor would find such a journal an almost irresistible target.

REFERENCES
1 *The Analyst*, Vol. 69, No. 816, p. 97, 1944.
2 Bulloch, *History of Bacteriology*, OUP, 1938, p. 53.
3 *Liebig's Annalen der Chemie*, Vol. 29, p. 100, 1839.
4 *Liebig's Annalen der Chemie*, Vol. 33, p. 308, 1840.
5 *Berichte der Deutschen Chemische Gesellschaft*, Vol. 19, pp. 3517–3568, 1886.
6 Sir John Read, *Humour and Humanism in Chemistry*, Bell, 1947, p. 216.
7 *Physics Today*, Vol. 19, No. 8, p. 31, 1966.
8 *Phoenix* (Imperial College, London) Autumn 1957.
9 *Close-up* (Armstrong-Whitworth and Jarrow Metal Industries, Ltd.) February 1958.
10 *Shipbuilding and Shipping Record*, Vol. 91, p. 280, 1958.
11 *Shipbuilding and Shipping Record*, Vol. 91, p. 301, 1958.
12 *Proceedings of the Chemical Society*, p. 383, 1959.
13 C D Graham, Jr. *Metal Progress*, Vol. 71, No. 5, p. 75, 1957.
14 *The Scientist Speculates* (an anthology of partly-baked ideas), Editor I J Good, Heinemann, 1962, p. 52.
15 *Proceedings of the Chemical Society*, p. 173, 1960.
16 *The Scientist Speculates*, p. 208.
17 Paul Junnings, *Oddly Enough*, Reinhardt, London, 1950, p. 147.
18 —*Even Oddlier*, Reinhardt, London, 1952, p. 148.
19 S Holmes, quoted by J Watson and A C Doyle: *The Adventure of the Bruce-Partington Plans*.
20 —*The Sign of Four*.
21 *Journal of Chemical Education*, Vol. 22, p. 509, 1945.
22 *Journal of Chemical Physics*, Vol. 25, No. 3, p. 606, 1956.
23 A C Doyle, *The Return of Sherlock Holmes*.
24 Pimentel and McClellan, *The Hydrogen Bond*, W H Freeman, 1960, p. 237.
25 M P Cava and D R Napier, *Journal of the American Chemical Society*, Vol. 78, p. 500, 1955.
26 *Non-Benzenoid Aromatic Compounds*, Editor D Ginsberg, Interscience Publishers, 1959, p. 57.
27 *Nature*, Vol. 230, No. 4941, p. 137, 1964.
28 *New Scientist*, Vol. 27, p. 37.
29 *IBM Journal of Research and Development*, Vol. 2, p. 282, 1958.

That thing will never work

DON GROVES

Condensed from
DAC News, Detroit
Athletic Club,
(March 1972)
pp 19–22.

These predictions illustrate the fact that there is no period in history when further improvement must end.

A committee of the British Parliament in 1878 reported Thomas A Edison's ideas of developing an incandescent lamp to be 'good enough for our transatlantic friends . . . but unworthy of the attention of practical or scientific men.'

A good navy man, Rear Admiral George W Melville, had this to say in 1901 about the possibility of building a successful airplane: 'Outside of the proven impossible, there probably can be found no better example of the speculative tendency carrying man to the verge of the chimerical than in his attempts to imitate the birds, or no field where so much inventive seed has been sown with so little return as in the attempts of man to fly successfully through the air . . . there is no basis for the ardent hopes and positive statements made as to the safe and successful use of the dirigible balloon or flying machine, or both, for commercial transportation or as weapons of war . . ..'

Even *after* the invention of the airplane, the famous astronomer William H Pickering, said with regard to airflight that: '. . . The popular mind often pictures gigantic flying machines speeding across the Atlantic and carrying innumerable passengers in a way analogous to our modern steamships . . .. Such ideas must be wholly visionary, and even if a machine could get across with one or two passengers, the expense would be prohibitive to any but that capitalist who could own his own yacht.

'Another popular fallacy is to expect enormous speed to be obtained. It must be remembered that the resistance of the air increases as the square of the speed and the work as the cube . . .. If with 30 hp we can now attain a speed of 40 mph, then in order to reach a speed of 100 mph we must use a motor capable of 470 hp . . . it is clear that with our present devices there is no hope of competing for racing speed with either our locomotive or our automobiles.'

197

When Samuel F B Morse offered to sell his telegraph to the US Government for $100 000, the Postmaster General rejected the offer on the basis that: '. . . the operation of the telegraph between Washington and Baltimore had not satisfied him that under any rate of postage that could be adopted, its revenues could be made equal to its expenditures.'

Errors and myopia

From *Men Who Found Out* by Amabel Williams-Ellis (New York: Coward-McCann), 1930, p 75.

William Harvey's discovery of the circulation of the blood (1628) received this response as recorded by John Aubrey, a contemporary:

> . . . I heard Harvey say that after his book came out, he fell mightily in his practice. 'Twas believed by the vulgar that he was crack-brained, and all the physicians were against him. I knew several doctors in London that would not have given threepence for one of his medicines.

From *Makers of Science* by I B Hart (Oxford: Oxford University Press) 1923, p 243.

When Georg Simon Ohm published his theory of electricity in 1827, his book was called 'a web of naked fancies' and the German Minister of Education said that:

> . . . a physicist who professed such heresies was unworthy to teach science.

Ohm lost his teaching position.

From *Father of Radio, the Autobiography of Lee de Forest* (Wilcox and Follett), 1930, p 227.

Friends of Lee de Forest asked:

> Well then, of what possible use can your 'radio-telephone' be? It can't compare with the wire 'phone, you say, and it can't cover the distances that the wireless telegraph can cover. Then what the hell use is it anyway, Lee?

Dr Burbeck, of the Mechanics Institute, England, gave this opinion, in 1839:

> The electric telegraph, if successful, would be an unmixed evil to society; would be used only by stock jobbers and speculators—and the present Post Office is all the public utility requires.

To offset such predictions there have been in all ages optimistic visionary predictions:

> The great bird will take its first flight on the back of the great bird, filling the world with stupor and all writings with renown, and bringing glory to the nest where it was born.

LEONARDO DA VINCI, 1505

Konstantin Eduardovich Tsiolkovsky (1857–1935) who anticipated many of the problems of modern rockets and their solutions, wrote:

> Earth is the cradle of the mind, but one cannot live in the cradle forever.

'Frankly, I don't like the look of the weather . . .'
(Courtesy of Norris—*The Vancouver Sun*.)

Extracts from *Modern Geography* *c*1740

Section IV—Containing some amazing geographical paradoxes

These are the chief Paradoxical Positions in Matters of Geography, which mainly depend on a thorough Knowledge of the Globe; and though it is highly probable, that they will appear to some as the greatest of Fables; yet we may boldly affirm that they are not only equally certain with the aforesaid Theorems, but also we are well assured, that there is no Mathematical Demonstration of Euclid more infallibly true in itself than every one of them. However, we think it not fit to pull off the Vizor, or expose those marked Truths to public View; since to endeavour the unmasking of them, may prove a private Diversion, both pleasant and useful to the ingenious Reader, at his more vacant Hours . . .

Par. 3 There is a certain Place of the Earth, at which if two Men should chance to meet, one would stand upright upon the Soles of the other's Feet, and neither of them would feel the other's weight, and yet they both should retain their natural Posture.

Par. 4 There is a certain Place of the Earth, where a Fire being made, neither Flame nor Smoke would ascend, but move circularly about the Fire. Moreover, if in that Place one should fix a smooth or plain Table, without any Ledges whatsoever, and pour thereon a large quantity of Water, not one drop thereof would run over the said Table, but would raise itself up in a Heap.

Par. 8 There is a certain Island in the Aegean Sea, upon which, if two Children were brought forth at the same instant of Time, and living together for several Years, should both expire on the same Day, yea, at the same Hour and Minute of that day, yet the Life of one would surpass the Life of the other by divers Months.

Par. 11 There is a certain Hill in the South of Bohemia, on whose Top, if an Equinoctial Sun-dial be duly erected, a Man that is Stone-blind may know the Hour of the Day by the same, if the Sun shines.

Par. 42 There are two distinct places on the Continent of Europe, so situated, in respect of one another, that though the first doth lye East from the second, yet the second is not West from the first.

Continental drip

ORMONDE DE KAY, JR, *Fellow of the Royal Geographical Society* †

Condensed from
Horizon (Winter
1973) 118–119.

[*Geophysicists of the world, are you ready? Here, hard on the heels of drift, comes another earth-shaking new theory derived from simply looking at maps.*]

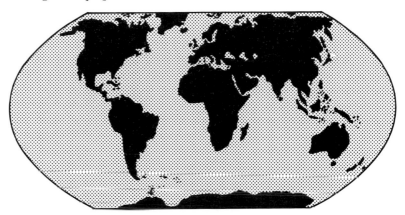

In the 1950s and 1960s came the first concerted international effort to study the Earth as a whole. It was found that the ocean floors are traversed by mountain ranges thousands of miles long. These oceanic ridges are zones of frequent earthquakes; periodically magma spews forth from fissures in them to spill down their slopes and widen the seabeds on either side. Scientists perceived that the lines of seismic activity along the oceanic ridges—and along the great circum-pacific earthquake belt as well as along certain earthquake belts extending overland—mark, in effect, the boundaries of immense 'plates' on which the continents 'drift'. Today the once-despised theory of continental drift is accepted everywhere as fact.

Well, I, too, have been looking at maps, and I, too, have been struck by a revelation no less global in scope. From it I have constructed a hypothesis, which I call, provisionally, the theory of continental drip.

As the term suggests, continental drip is the tendency of land masses to drip, droop, sag, depend, or hang down—like wet paint in the Sherwin–Williams trademark—except that they cling to the Earth's surface below the equator instead of falling off into space. The cartographic evidence for drip is at

† In arrears

least as impressive as that for drift. It is, indeed, as plain as the nose on your face—and probably more pendulous.

Let's look at the World map. Africa and South America—those very land masses, by coincidence, whose congruent coasts first inspired both Bacon and Wegener to their wild surmise [of continental drift]—are textbook examples of drip, with their broad tops and tapering lower extremities. But so is North America, with Baja California and Florida dangling down at its sides; Greenland, too, clearly shows the characteristics of drip.

Of Europe's major projections, all except Jutland hang down, but with a difference: Scandinavia and Iberia droop south*west*, while Italy and Greece droop south*east*. These departures from the vertical are caused by continental drift: Eurasia is drifting eastward, dragging the first two peninsulas behind it. As for the second two, they are paralleled by two earthquake zones—'plate boundaries'—that extend, respectively, between the Alps and the Aegean and between the eastern end of the Black Sea and the northwestern corner of Iran. The Crimea, incidentally, is another example of continental drip.

In Asia we find a consistent record of drippage. From its southern flank droop pendulous India, then the spindly Malay Peninsula, and finally, sagging Indochina.

Australia appears to defy the laws of continental drip. (This strange land, we may recall, once seemed to defy the laws of zoology and botany as well.) But since Australia was the last continent to assert its independent identity, still having been attached to Antarctica a mere forty million years ago, it may simply not have been around long enough to show the more pronounced effects of drip.

By the assumptions of drip theory we would expect earthquake-free islands to extend, as a rule, north–south, and to lie south of large land masses from which they have dripped. Again the map bears out our expectations: we find Madagascar below the horn of Africa, Ceylon below India, Tasmania below Australia, Borneo below Southeast Asia, and Hainan below the bulge of China.

The British Isles were once, as we know, linked by broad land bridges to each other and to Europe. Here, again, however, their origin would seem less important than their north–south orientation, caused by drip.

The shorelines ringing the Arctic Ocean are broken and irregular, but no arm of land projects markedly north. Antarctica's chief significance to drip theory is simply that it is where it is, as far south as it can be.

If anyone still questions the reality of continental drip, let him the turn the World map upside down. The land masses now shoot upward in a giddy profusion of spikes, spires, spindles, knobs, and domes, while the earthquake-free islands rise like captive balloons from their vertiginous 'skyline'. This global show of energetic upward movement confirms the impression of languid downward drip received earlier.

Drift and drip are complementary; while the first determines the position of land masses, the second helps determine their shape.

A vexing question remains: What causes continental drip? A few possible explanations come to mind: some palaeomagnetic force, for example, unsuspected and therefore undetected, centered in massive, mountainous Antarctica and perpetually tugging at the lower hems of land masses. Or drip might somehow be the result of the Earth's rotation, or of lunar attraction. One conclusion, however, would seem inescapable: contrary to the teachings of science, but as every schoolchild has always known, north really is up, and south down!

Sir Isaac Newton

Truth is ever to be found in simplicity, and not in the multiplicity and confusion of things . . . He is the God of order and not of confusion.

Acknowledgments

The Institute of Physics and the Compiler gratefully acknowledge permission to reproduce copyright material listed below. Every effort has been made to trace copyright ownership and to give credit to copyright owners but if, inadvertently, any mistake or omission has occurred, full apologies are herewith tendered.

American Association for the Advancement of Science: from *Science*. Copyright by the American Association for the Advancement of Science.

American Association of University Professors: from *AAUP Bulletin*.

American Institute of Physics: from *Applied Optics, Physics Today, American Journal of Physics, American Physics Teacher, Physics Teacher, Review of Modern Physics, Journal of the Acoustical Society of America*.

American Scientist: from *The Legal Value of π, and Some Related Mathematical Anomalies* by M H Greenblat, December 1965. Reprinted by permission of *American Scientist*, journal of Sigma Xi, The Scientific Research Society.

The Archimedeans (Cambridge University Mathematical Society): poem from *Eureka*, number 17, October 1954, pp 5–7. Reprinted by permission of The Archimedeans, The Arts School, Bene't Street, Cambridge.

Isaac Asimov: from *Treasury of Humour*. By permission of the Woburn Press, London.

Marque Bagshaw: *The Quatorze Connection*.

Basic Books, Inc and Andre Deutsch: from *The First Three Minutes: A Modern View of the Origin of the Universe* by Steven Weinberg. Copyright © 1977 by Steven Weinberg.

Bell & Hyman Ltd: from *Humour and Humanism in Chemistry*.

Basil Blackwell Publisher Limited: from *Oxford Outside the Guidebooks*.

Felix Bloch and the American Institute of Physics: from a talk given at the Washington DC meeting of The American Physical Society 26 April 1976.

Bobbs-Merrill Educational Publishing: from *Zeno's Paradoxes* edited by Wesley Salmon, copyright © 1970.

Elizabeth (Wood) Bogert: *Relativity* by R W Wood.

Alexander Calandra: *Free Thinking*.

Joel E Cohen: *Designer Genes*.

The Copyright Agency of the USSR: from *Novy Mir*.

The Daily Telegraph: On the Spot.

Nuel Pharr Davis: from *Lawrence and Oppenheimer*.

Dover Publications, Inc: from *How to tell the Birds from the Flowers—and other Wood-cuts* by R W Wood.

Ronald D Edge: *My Anecdotage: Apocryphal Tales*.

Sir Brian Pippard: from *Post-prandial Proceedings of the Cavendish Society*.

Poem: reprinted by permission from *Nature*, volume 218, p. 797. Copyright © 1966 Macmillan Journals Limited.

J B Priestley: from *Man and Time*, copyright © 1964 by Aldus Books Limited, London.

Punch: quotation.

Punch and *The Observatory* Magazine: *The Satellite Question*.

The Rank Order on Campus by Pearl Aldrich. Reprinted with permission. Copyright © 1981 by The Chronicle of Higher Education, Inc. Appeared originally in *The Chronicle of Higher Education*, 14 January 1980.

Eric M Rogers: *Demon Theory of Friction*.

Ian Rose: from *Fellowshipmanship*. Originally published in *Canadian Medical Association Journal*, volume 87, 8 December 1962.

David W Roubik: from *Science*.

Royal Astronomical Society Club: *The Einstein and the Eddington*.

Royal Society of Chemistry: quatrain.

Royal Statistical Society, J M Chambers and A M Herzberg: *A Note on The Game of Refereeing*.

Rutherford Appleton Laboratory: from *Orbit*.

George H Scherr: from *The Journal of Irreproducible Results*.

Charles Scribner's Sons: from *The Works of Robert Louis Stevenson*, 1925.

Servire: from *De Sterrekunde en de Mensheid* by M Minnaert, Servire B V, Katwijk, 1946.

John N Shive: *Physics Instructor*.

Norman Sperling: *Saving Time*.

Donald R Stauffer: *Stauffer's Law of the Entropy of Change*.

George Steiner: from *A Future Literacy*. Copyright © 1971 by George Steiner.

Margaret W Stubbs: from *The Space Child's Mother Goose* by Frederick Winsor, Simon & Schuster, 1958.

A W S Tarrant: *The Use of Small Dogs in Physics Teaching*.

Taylor & Francis, Ltd: from *Philosophical Magazine*.

Teaching: from *Life with Picasso* by F Galot and C Lake. Copyright © 1965. Used with permission of McGraw-Hill Book Company.

The University of Chicago Press: *Hypothesis* by A S Parkes, from *Perspectives in Biology and Medicine*, 7, 1958; and the Foreword by John Ziman to *Originality and Competition in Science* by Jerry Gaston, 1973. Reprinted by permission of The University of Chicago Press.

The Vancouver Sun and Len Norris: cartoon.

The Washington Post: The Dreistein Case by J Lincoln Paine (Arnold Kramish), originally published in *The Washington Star*.

Susan Watkin: *Extracts from Modern Geography c 1740.*
Wilcox and Follett: from *Father of Radio* by Lee de Forest.
John Wiley & Sons, Inc: from *A Biographical Dictionary of Scientists* and from *Foundations of Physics.*
Amabel Williams-Ellis: from *Men Who Found Out.*
Jerry D Wilson and Harold Knox: *Error Bars.*
Susan Winarchick Cohick, Laura Kate Hutton, Janet Briggs and the Society of Physics Students, The Pennsylvania State University: *All Power to Particles.*